BLAISE PASCAL

Also by Donald Adamson

THE GENESIS OF *LE COUSIN PONS*

DUSTY HERITAGE

THE HOUSE OF NELL GWYN

BALZAC: *Illusions Perdues*

LES ROMANTIQUES FRANÇAIS DEVANT LA PEINTURE ESPAGNOLE

Translations of Balzac:
LA RABOUILLEUSE
URSULE MIROUËT

and of Maupassant:
STORIES

Blaise Pascal

Mathematician, Physicist and Thinker about God

Donald Adamson

'Qu'on ne dise pas que je n'ai rien dit de nouveau, la disposition des matières est nouvelle. Quand on joue à la paume, c'est une même balle dont joue l'un et l'autre, mais l'un la place mieux'
(Thought 696)*

'Let no one say that I have said nothing new; the arrangement of the material is new. When playing tennis, both players use the same ball, but one plays it better'

M
St. Martin's Press

First published in Great Britain 1995 by
MACMILLAN PRESS LTD
Houndmills, Basingstoke, Hampshire RG21 2XS
and London
Companies and representatives
throughout the world

A catalogue record for this book is available
from the British Library.

ISBN 0–333–55036–6

10 9 8 7 6 5 4 3 2 1
04 03 02 01 00 99 98 97 96 95

Printed and bound in Great Britain by
Antony Rowe Ltd
Chippenham, Wiltshire

First published in the United States of America 1995 by
Scholarly and Reference Division,
ST. MARTIN'S PRESS, INC.,
175 Fifth Avenue,
New York, N.Y. 10010

ISBN 0–312–12502–X

Library of Congress Cataloging-in-Publication Data
Adamson, Donald.
Blaise Pascal : mathematician, physicist and thinker about God /
Donald Adamson.
p. cm.
Includes bibliographical references and index.
ISBN 0–312–12502–X
1. Pascal, Blaise, 1623–1662. 2. Science—History.
3. Philosophy—History. 4. Scientist—France—Biography.
5. Philosophers—France—Biography. I. Title.
Q143.P17A33 1995
194'.092—dc20 94–31137
CIP

To the memory of my beloved mother

Contents

List of Illustrations

List of Figures

Acknowledgments

For permission to use photographic material I am grateful to the Niedersächsische Landesbibliothek, Hanover; the Staatlicher Mathematisch-Physikalischer Salon, Dresden; the British Library, London; and the Principal and Fellows of Newnham College, Cambridge.

I also acknowledge with gratitude the assistance kindly extended to me by the following institutions and persons: the Science Reference and Information Service of the British Library; the library of Newnham College, Cambridge (its present and former librarians: Ms Deborah Hodder and Miss Ann Phillips); the University Library, Cambridge; Wolfson College, Cambridge; the Bibliothèque Municipale & Universitaire de Clermont-Ferrand; the Musée d'Histoire et d'Art local (Musée du Ranquet), Clermont-Ferrand; the Leibniz-Archiv, Hanover (keeper: Dr Herbert Breger); St Deiniol's Library, Hawarden, Clwyd; IBM Europe SA (Mr Alan Gillings); the Japan Association for Philosophy of Science; the British Museum, London (Mr John Leopold and Mr David Thompson); Dr Williams's Library, London (in particular, Mrs Janet Barnes); the Imperial College & Science Museum Libraries (library and information service of the National Museum of Science & Industry); the London Library; the National Art Library (library of the Victoria and Albert Museum, London); the Royal Society, London (archivist: Ms Mary Sampson); Sion College Theological Library, London; the University of London Library; the Bodleian Library, Oxford; the Bibliothèque Nationale, Paris; the Bibliothèque Sainte-Geneviève, Paris; the Université de Paris-Sorbonne (Paris IV); the Musée National des Granges de Port-Royal, Port-Royal des Champs, Yvelines (curator: M. Philippe Le Leyzour); the Musée National des Châteaux de Versailles et de Trianon (M. Roland Bossard); Mr Leslie Bedford; Mr David Birch; Dr Terry Boddington; the Reverend Professor Henry Chadwick; Mr John Darby; Dr Wolfgang Dittrich; Ms Evelyne Draper; Viscount Eccles; Dr John Gibbins; Professor Brian Gray; Frau Anke Hölzer; M. Patrick Jeanne; M. Ernest Jousset; Dr Klaus Schillinger; Mrs Paula Scott-James; Professor Thomas Whiteside; and my wife.

To the British Library I am most deeply indebted, however, not only for the almost incomparable richness of its collections but also for the helpfulness and courtesy which I have always been shown there.

DONALD ADAMSON

1
Introduction

Blaise Pascal was born on 19 June 1623 at Clermont (in 1630 the city was to be amalgamated with Montferrand and become Clermont-Ferrand): he was the third child and only son of Étienne Pascal's four children; their mother died in 1626. Clermont was a city steeped in the history of Christendom: at the conclusion of an ecclesiastical council held there in 1095, Pope Urban II had blessed the departure of the First Crusade.

Pascal's father's and mother's families had long been established in Auvergne.[1] The roots of his paternal grandmother's family of Pascal can be traced back as far as 1443: originally resident at Le Pertuis, they were granted armorial bearings in 1480. Blaise's paternal grandfather Martin Pascal was a taxation commissioner for Clermont who later rose to be private secretary to the wife of King Henri III and in 1587 became Treasurer of France for the generality of Auvergne at Riom. His father had been a presiding judge at the taxation court in Montferrand. Later, after a premature period of early retirement in Paris, he moved to Rouen where he became a senior commissioner of taxes.

From 1632 until 1639 the young Pascal lived with his family in Paris but did not attend any school or university, being principally educated by his father, who in turn had been educated by his father (in Greek, Latin, mathematics, history, philosophy, theology and canon and civil law: 45): Étienne thus became a man of wide scientific and mathematical interests, and the inventor of what is known as Pascal's *limaçon* (i.e., the conchoid of a circle with respect to one of its points, which he applied to the problem of trisecting an angle). The 'rigor and originality' of the boy's education, writes C.M. Cox,[2] 'can be likened only to the discipline of John Stuart Mill'.[3] He is said, at the age of eleven, to have produced a 'Treatise on Sounds', since lost (4).

Pascal was also strongly influenced by the informal scientific discussion group initially presided over by the elderly Minim friar Marin Mersenne, one of the most gifted and enquiring men of his age. In 1639,[4] and again (after his return to Paris) from 1648 onwards (**I 169**), he would accompany his father to some of the regular Saturday meetings of this group; here he met Roberval, Le Pailleur,[5] Mylon, Carcavi,[6] Auzout, Mydorge and sometimes Desargues and Gassendi. He was not only a studious boy but also, in mathematics, a child prodigy. Without help or instruction from anyone he mastered the primary elements of Euclid (4–5), and tradition has it[7] that even as a boy he devised one, or more than one, method of demonstrating, without fully proving, that the sum of the angles of a plane triangle is exactly equal to two right angles.[8] At the age of fourteen or fifteen he became a fervent admirer of the geometrician Gérard Desargues. Prior to publishing an *Essay on Conic Sections* in February 1640,[9] when sixteen years of age, he had already – about June 1639 – discovered the theorem of the Mystic Hexagon ('Pascal's Theorem').

From *circa* December 1639 until May 1647 Pascal lived with his father and sisters in Rouen as Étienne Pascal had been appointed to assess and collect taxes in upper Normandy.[10]

The basic designing and production of a calculator for his father's use in the course of these duties took Blaise almost three years, from 1642 to 1645, during which time (in the intervals of his illness) he seems to have devoted all his energies to the task. In 1645, in an open letter to the Chancellor of France, Pierre Séguier (349–53), he outlined its advantages and methods of operation. With one possible exception, he became the first person ever to manufacture and market a desktop mechanical calculator. Even so, they were unwieldy and fairly expensive products, most of them costing about 100 *livres*,[11] i.e., approximately £500 at present values: the price of a modern desktop computer. By 1652 fifty prototypes had been produced, but few machines were sold,[12] and manufacture of Pascal's arithmetical calculator ceased in that year.

So ended the second phase of his scientific career.

Pascal had been brought up by his father in a strict religious atmosphere characterized by regular prayer and frequent church-going. In 1646 Étienne Pascal, having injured his thigh, was tended at home by two young brothers who were devout followers of the parish priest of Rouville, just outside Rouen: Rouvillism was a religious movement strongly influenced, in its fervour and intensity, by the Jansenists Arnauld and Saint-Cyran. Daily contact with these young men deepened the family's religious commitment; this experience is sometimes known as Pascal's 'first conversion'. The

Figure 1.1 Pascal's Theorem of the Mystic Hexagon

Pascal showed that if a hexagon is inscribed within a conic, the three points of intersection of opposite pairs of sides (AB/EF, BD/CE, AC/DF) of that hexagon will always lie on a straight line GH.

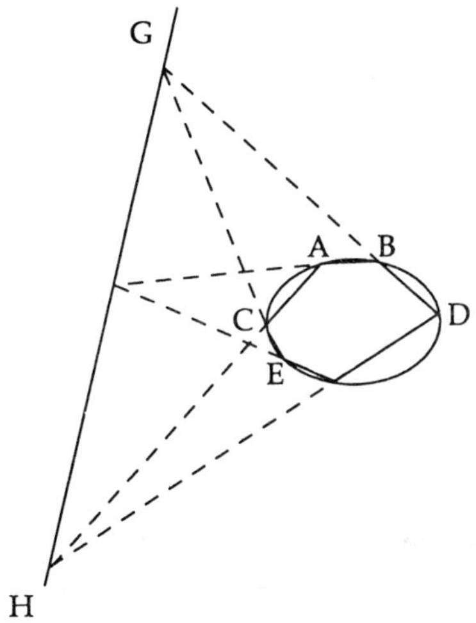

Pascal's own formulation of the Mystic Hexagon was apparently more complex.

From notes taken by E.W. von Tschirnhaus and Leibniz (and now at the Niedersächsische Landesbibliothek, Hanover), it would seem that a double hexagon is inscribed within a conic.

The theorem of the Mystic Hexagon does not feature in the *Essay on Conic Sections*.

young men's influence was greatest upon Blaise and his younger sister Jacqueline, who took the veil in May 1652 and made her final profession in June 1653; he had objected to her entering upon the religious life, even to the extent of refusing for a whole year to pay the 'dowry' which the convent of Port-Royal expected of all its professed nuns.

In January–March 1647 Pascal and two friends engaged in fierce debate with the parish priest, and disaffected Capuchin friar, Jacques Forton about the nature of the Holy Trinity and the causality of the world. This dispute led to the downfall of the Archbishop of Rouen's coadjutor, Jean-Pierre Camus, Bishop of Belley,[13] and to Forton's removal from his benefice on grounds of heresy.

At Rouen, Clermont-Ferrand and in Paris, between October 1646 and September 1648, and from then on until March 1651, Pascal either conducted or caused to be conducted various experiments on atmospheric pressure and the existence of vacuums.[14] The most important of these experiments was carried out on 19 September 1648 by his brother-in-law Florin Périer on the Puy de Dôme; another notable one was conducted from the tower of the church of Saint-Jacques de la Boucherie (399). Gassendi has also described[15] how Pascal carried a gradually distending balloon, made from a carp's bladder, to the very top of the Puy de Dôme, thus satisfying himself that air was a compressible substance.

Although he did not discover the actual principle of the barometer[16] (credit for which belongs to Torricelli), Pascal supplied, as early as 1647, virtually incontrovertible proof that vacuums existed. He also showed that the height of the column of mercury in a barometer decreases as it is carried upwards through the atmosphere. From this discovery it followed that a vacuum existed above the atmosphere, thus contradicting Descartes's denial of the existence of vacuums and his contention that all space is filled with matter.[17]

On 23 and 24 September 1647 Pascal received visits from Descartes. These, it seems, were Pascal's only encounters with the philosopher who was already renowned both for his *Discourse on Method* and for his analytical or co-ordinate methods of applying algebra to geometry. They were much less happy occasions than the meetings between Spinoza and Leibniz in 1676, though Leibniz later played down the closeness and friendliness of those exchanges. Descartes admired the mechanical calculator but, disbelieving in the existence of vacuums, engaged in heated argument with Pascal on that subject. In the following month (4 October 1647) Pascal reported some of his recent scientific activities in the pamphlet *New Experiments Concerning Vacuums.* Soon after the Forton dispute Pascal's atmospheric experiments led to a second but only slightly less acrimonious controversy with the Jesuit (and neo-Aristotelian) priest Étienne Noël (October–November 1647;[18] March or April 1648)[19] in which Noël, though accepting the theory of atmospheric pressure, denied the existence of vacuums whilst Pascal defended both: in the second case with perhaps undue vigour as he himself had been ready, in *New Experiments Concerning Vacuums,* to accept the notion of Nature's limited abhorrence of the void (368). He published a further account of this scientific activity (*Narrative Account of the Great Experiment on the Equilibrium of Liquids*) in October 1648.

Pascal's achievements in physics entitle him to be regarded as one of the founders of hydrostatics and hydrodynamics.

Not long after his father's death, which occurred on 24 September 1651, Pascal wrote to his sister and brother-in-law Gilberte and Florin Périer about the meaning of death for the Christian believer (490–501: 17 October 1651).

From May 1653 until October 1654 he was almost constantly engaged in scientific enquiry, both in mathematics and in physics; he again attended meetings of his scientific discussion group (**III 431**),[20] now chaired by François Le Pailleur and perhaps occasionally by Claude Mylon. In his *Treatise on the Equilibrium of Liquids* (*circa* 1653) he formulated what came to be known as Pascal's Principle, or law, of pressure. This treatise is a complete outline of a system of hydrostatics, the first in the history of science; it embodies his most distinctive and important contribution to physical theory. The results of his early work on the void were now incorporated into a theory of the statics of fluids. Dealing with the effects of the weight of the atmosphere largely in terms of the principles enunciated in the earlier treatise, he went on to write a *Treatise on the Weight of the Air Mass* (*circa* 1654) which, when finally published, led to the measurement of altitude from barometric pressure[21] and to many further inventions. Pascal also put the finishing touches to treatises on geometry (e.g., *The Generation of Conic Sections,* essentially completed by 1648), and in his *Treatise on the Arithmetical Triangle* conducted important research into combinatorial analysis. He also studied probability theory (in his celebrated correspondence with Pierre de Fermat, June/July–25 September 1654), the factorization of multiples and the summing of powers of series of numbers in arithmetical progression.

So ended the fourth period of Pascal's scientific activity, particularly important in that the Fermat correspondence and the *Treatise on the Arithmetical Triangle* have been the foundations of the science of statistics – with all the many benefits (and perhaps a few disadvantages!) which that science has conferred upon modern life.

In October 1654, however, Pascal experienced some sort of revulsion from mathematics, writing to Fermat, with much more than a touch of irony, on the 27th of that month (**III 431**) that the latter's latest discoveries in probability theory were now beyond his understanding.

It was in this same year, 1654, that Pascal's so-called 'accident on Neuilly bridge' (1365) may have occurred.[22] According to this story, he was driving across the bridge with friends when the horses of their carriage suddenly took fright, bolting over the parapet of the bridge and plunging into the Seine; the carriage itself remained precariously perched on the parapet, suspended as it were in space, until its occupants were eventually rescued. This episode is said to have had a momentous effect upon Pascal, persuading him to withdraw from the world and live entirely alone. It was, according to Kierkegaard, one of those decisive experiences – like Saul's encounter on the road to Damascus,[23] or Luther walking beside his friend struck dead by lightning, or Claudel's vesper conversion in Notre-Dame on Christmas Day 1886[24] – which 'shatter a man without killing him'.[25] Such experiences, claims Kierkegaard, propel a man irresistibly into lifelong 'service of the Absolute'. They are, he says, 'the infinite intensely concentrated in a single pressure and in a single moment of time'. Voltaire, on the other hand, stressed this 'accident on Neuilly bridge' as a way of undermining the validity of Pascal's subsequent religious conversion.[26]

On 23 November of the same year (1654) he underwent this 'night of fire', a mystical experience sometimes known as his 'second conversion', in which he affirmed beyond any doubt his deep Christian belief, also committing himself heart and soul to a life of Christian self-sacrifice. A record of this visionary experience is the *Memorial* (913*), a note written very shortly after its occurrence, two copies of which Pascal wore sewn into the lining of his doublet – and transferred from garment to garment – until the day of his death.[27] It ends with a quotation from Psalm CXIX 16: *I will not forget Thy word.*

Now began his visits to Port-Royal des Champs, some twenty-three miles south-west of Paris; it was, all in one, a convent, semi-monastic community and school of Jansenist leanings. Altogether he seems to have paid no more than five visits to this establishment, and probably as few as three or four.[28] Particularly influential upon him during these visits were Nicole, Singlin and his newly appointed spiritual director Isaac Le Maître de Saci, all of them priests and all Jansenists. The *Conversation with M. de Saci Concerning Epictetus and Montaigne* is the first fruit of Pascal's association with Port-Royal. It is the record of one or more lengthy discussions with M. de Saci, and a rebuttal both of stoicism and of scepticism and of all that is secular in man's attitude towards the world. Whilst at Port-Royal des Champs in January 1655, Pascal may also have found time to compose *The Mystery of Jesus* (919*), a contemplation of the Passion of Our Lord. About this time he may have drafted *On the Conversion of the Sinner*,[29] whose author expresses a sense of the futility of worldly things. And probably later in 1655 he drew up a detailed *Summary of the Life of Jesus Christ* under 354 headings. Also in that year, or thereabouts, he pondered the origins and growth of the Early Church. His thoughts on this subject are set down in the *Comparison between Christians of Early Times and Those of Today*.[30]

Between January 1656 and March 1657 Pascal published eighteen *Provincial Letters*, pseudonymously, under the name of Louis de Montalte. A defence of the fundamental principles of Jansenism against the laxities of the Jesuits, they were his main written excursion into religious controversy. Few people suspected him of being the author of these *Letters*, which dealt an exceedingly savage blow to the Society of Jesus, especially condemning their moral teaching and their attitude towards penitents in the confessional. Pascal may well have been visualizing his brother-in-law Florin Périer as he addressed the *Provincial Letters* to his imaginary correspondent.

On 24 March 1656 the so-called Miracle of the Holy Thorn occurred to his niece and goddaughter, the ten-year-old Marguerite Périer, who for three years had been suffering from a fistula of the eye. This led him, about September 1656, to regard miracles as a crucial feature of the defence of the Christian religion which by then he had in mind. His nine (extant) *Letters to Mlle de Roannez*, composed between September and December 1656, are a series of religious meditations, contemplative in tone. They stress the unity of the Church, the Hidden God, and the reflection of an eternal Reality in all earthly things.

Pascal's *Writings on Grace*, unpublished until 1779, dealt with a subject previously treated in the *Provincial Letters*, the nature and operation of divine grace and the meaning of the sacrament of Holy Baptism. Having become more independent in his religious outlook, he was now exploring these arcane matters for himself rather than pitting the Augustinian outlook against that of both Dominicans and Jesuits, as had been the case in the first, second and third of the *Provincial Letters*.

During his later years – perhaps from about 1656 onwards – he would sometimes wear a spiked iron belt next to his body (13–14), pressing these spikes into his flesh if he felt he was deriving any vanity or even actual pleasure from his conversations with visitors. And when racked by pain and sickness during his last terrible four years of terminal illness (12–13, 19), he would practise the same self-mortification in an attempt to rouse himself from his lethargy. This spiked belt remained a closely guarded secret, not even shared during his lifetime with members of his family.

In the summer of 1656, and again in March 1657, he was still engaged with Fermat and Carcavi in investigations of probability theory. During the autumn or winter of 1657,[31] at the request of Arnauld, Nicole and perhaps other Solitaries of Port-Royal, he helped to draw up a 'Port-Royal Grammar'. This was a teaching aid intended for the schoolmasters at that establishment; it exerted a powerful influence upon the textbook eventually published two years before his death; Chapter VI of the so-called *Port-Royal Grammar*[32] is his own work (1455).[33] It suggested a new reading method based upon the phonetic value of vowels and consonants: *m*, for example, would be pronounced *mm* and not *em*; likewise, *o* and the French, though not the Italian, *au* would be given the same pronunciation. This, it may be thought, was an obvious enough innovation but it was not considered to be so in 1655 and even Pascal's sister Jacqueline raised objections to it (1455).

About this time Pascal also drafted the outlines of 'Elementary Geometry', a textbook which in the long run was to have greater repercussions than the 'Port-Royal Grammar'. Its few surviving passages mainly consist of the *Introduction to Geometry*, a fragment preserved by Leibniz (602–4), and the all-important remarks about definitions which occur not only in *On the Geometrical Mind* (576–92) and *On the Art of Persuasion* (592–602) but also, in modified form, in

Arnauld's *New Elementary Geometry*[34] and even to some extent in Book I of the *Port-Royal Logic.*[35] As mathematics, this work of popularization is notable for its straightforward exposition of Euclidean theorems in terms of projective geometry.

The *Port-Royal Logic,*[36] a collaborative work overseen by Arnauld and Nicole, remained for almost two centuries the basic source-book for the study and teaching of logic. Its statements about definitions were the very core of the book: they remained unchallenged, and substantially unmodified, until the writings of Augustus De Morgan and George Boole; indeed, the *Port-Royal Logic* was still in use, both on the Continent and in the British Isles,[37] until just over a hundred years ago. In Book I Pascal introduces into European thought the notion of 'termes primitifs' (or undefined terms), observing that these, together with various unproved propositions, are the foundation of the experimental sciences. He – and Arnauld and Nicole after him – insist, however, that for the sake of clarity of thought and discussion such 'undefined terms' should be completely understood: not for him the tautological muddle of Étienne Noël's definition of light (1441), wickedly transformed by Pascal into 'la *lumière* est un mouvement *luminaire* des corps lumineux' (377) in the thick of the dispute about vacuums! Pascal does not accept that 'undefined terms' are as arbitrary as the nominal definitions (*définitions de nom*) which, to his mind, are self-referential. For, as he observes elsewhere (380–1),[38] it is no less easy for such 'nominal definitions' to refer to an impossibility than it is for them to refer to anything actually existent.

During the years that followed the writing of the *Provincial Letters* and lesser religious works Pascal began to devote his life to the advancement of Christianity. At the heart of this activity were the *Thoughts,* which, begun around September 1656, date mainly from 1657 and 1658.[39] This is a largely unfinished work, and (in the view of many commentators) is the most ambitious of all attempts to convert the agnostic and the atheist to religious belief.[40]

Between January and July 1658 Pascal seems to have taken the leading part in drafting a deposition and a series of statements on behalf of the parish priests of Paris and perhaps also of other dioceses: these were the *Deposition on Behalf of the Parish Priests of Paris* (25 January 1658), the *Second Statement by the Parish Priests of Paris* (2 April 1658), the *Draft Mandamus Against the* Defence of the Casuists[41] (*circa* 7 June 1658), the *Fifth Statement by the Parish Priests of Paris* (11

June 1658), which in its author's own view[42] was 'the finest thing he ever wrote' (**VII 353**), and the *Sixth Statement by the Parish Priests of Paris* (24 July 1658). With casuistry as their target, these statements prolonged and extended the dispute with the Jesuits: Pascal's indictment of casuistry assumed a more indignant tone.

Returning yet again, and for the last time, to mathematics, he worked in that field from June 1658 until February 1659, his last such achievement being a *Letter to Carcavi* (341–3). In this fifth period of his mathematical involvement the focus of his attention was the cycloid, the curve formed by following a point on the circumference of a rolling circle; the fruits of his research were eventually to be embodied in four[43] *Letters from A. Dettonville to M. de Carcavi* (224–46) and in the annexes to those letters, the so-called *Treatises on Geometry* (247–312). Galileo, Torricelli, Descartes, Fermat and (above all) Roberval had already studied the properties of this curve; Roberval seems to have been the first mathematician to calculate the area of a cycloid by means of the method of indivisibles.[44] Carrying research into the cycloid an important stage further, Pascal now applied Cavalieri's calculus of indivisibles to the solutions of its quadrature, cubature, and centres of gravity. In particular, his resolution of the problem of the quadrature of the cycloid, pointing forward as it did to the infinitesimal calculus, was a mathematical achievement of the highest order.

Having made these discoveries, Pascal despatched his so-called *First Circular Letter Concerning the Cycloid* to Wren,[45] La Loubère, Wallis, Fermat, Huygens, Leibniz and other mathematicians of European renown, challenging them to produce solutions of two problems concerning the centres of gravity of a simple cycloid's solids of revolution. Pascal's *Second and Third Circular Letters Concerning the Cycloid* followed in July and October 1658 respectively. No mathematician, with the partial exceptions of Wallis and La Loubère,[46] came near to solving these difficult challenges, though it was whilst vainly attempting to solve them that Wren made his celebrated rectification of the general cycloidal arc. Pascal's *History of the Cycloid* (194–200), published within ten days of the deadline for receipt of entries to the competition, commented unfavourably on the submissions received from La Loubère and Wallis. But the *History of the Cycloid* was particularly controversial on account of its treatment of Torricelli. It is noteworthy, too, that the *History of the Cycloid* contains no reference to Cavalieri.

During the summer of 1658 Wren challenged Pascal and – as Seth Ward puts it[47] – other 'mathematicians in France' with the so-called 'Kepler's Problem': that of rectifying the general cycloidal arc. This was a problem to which Fermat, Roberval and the ailing Pascal all failed to provide a solution.

Amos Dettonville was Pascal's scientific alias from December 1658 until the end of all his secular intellectual work two months later in February 1659. It was an anagram of Louis de Montalte, the name he had used to disguise his identity when publishing the *Provincial Letters*.[48] Dettonville's *Sequel to the History of the Cycloid* (12 December 1658) and *Footnote to the Sequel to the History of the Cycloid* (20 January 1659) were even more scathing in their treatment of La Loubère. Not until his four (pseudonymous) *Letters from A. Dettonville to M. de Carcavi* (late in December 1658 and in January 1659) did Pascal propound the solutions to all the cycloid problems he had set. Between December 1658 and February 1659, still using the Dettonville alias, he published further mathematical discoveries, notably the properties of the solid of revolution generated by a spiral rotating around a cone; he also provided a method of determining the quadrature of simple, prolate and curtate cycloids.

The *Letters to Carcavi* are a Treatise on the Cycloid in all but name. A milestone in the development of the *infinitesimal* calculus, they led, but only indirectly, to the Newtonian *integral* calculus: it is unclear whether Newton himself was aware of his predecessor's achievements.[49] What is beyond doubt is that the *Letters to Carcavi* do not directly foreshadow Leibniz's discovery, in October 1675, of the *differential* calculus[50] (a discovery quite independent of Newton's formulation of the calculus some ten years previously).[51] Pascal did no more than devise an algorithm of the infinitesimal calculus, which was then still in its embryonic form of Cavalieri's method of indivisibles.[52] Stubbornly remaining within the disciplines of projective geometry,[53] refusing to employ Descartes's analytical methods and algebraic symbolism, and interpreting Cavalieri's theory of indivisibles in an arithmetical manner, he had not achieved the generalized formulations which both Leibniz and Newton were able to bring to bear upon the subject (the latter, by means of his method of fluxions).

Nevertheless, the understanding of conic sections – less so of prime numbers – had progressed very little by the time of the

Encyclopédie, a century or so later: such was the extent of his achievement in this area.

Early in 1659 Pascal met Henry Oldenburg during the latter's visit to Paris. England's equivalent to Pierre de Carcavi, Oldenburg was already to all intents and purposes secretary to the Royal Society of London as it was soon to become; he refers to Pascal as one of a group of French mathematicians which also included Roberval, Mylon and Clerselier.[54] During his last three and a half years, however (March 1659–August 1662), Pascal's commitment to science was slight. As his life drew to its close, he even tended to belittle the importance of scientific research (522–3).[55] During this last phase of his life his mental activity was directed elsewhere. Two years before his death (43), in what has been called a 'third conversion', he undertook a spiritual retreat. This experience, which may have lasted for several weeks (46), brought about a profound change in his style of life. Gone were all the luxuries of his existence: servants, carriage, horses, tapestries, silverware, best furniture, and his library[56] with the main exception of the Bible and St Augustine. Reduced to the strict necessities of life, he gave generously to the poor and went about Paris, from church to church, attending many religious services. His health was deteriorating to an alarming extent. He was totally debilitated from about April 1659 until the summer of 1660.[57] During this period of lethargy he may have composed a *Prayer to God Concerning the Proper Use of Illnesses*, a spiritual meditation of intense poignancy.

The closing years of Pascal's life were also embittered by disputes over the signing of the Formulary, imposed by Louis XIV in the name of Pope Alexander VII, which compelled French priests, monks and nuns to assent to the condemnation of Jansen's *Augustinus*. It is thought that in or about May 1661 Pascal had a hand in the writing of the *First Mandamus of the Vicars General Concerning the Signing of the Formulary Condemning* De Jure *Jansen's Five Propositions*. This distinguished between *de jure* and *de facto* condemnation and was something of a face-saving device: however, it enabled Jacqueline Pascal to sign the Formulary, although her remorse at having done so probably helped to bring on her early death (4 October 1661). Pascal came to believe that there should no longer be any compromise on this matter. Unlike many at Port-Royal des Champs, he was opposed to the signing of a Formulary in which no distinction was to be made between *de jure* and *de facto* condemnation (1075–7). These disputes led to a cooling of his friendship with Arnauld and Nicole.

During these closing years of his life he also won distinction both as businessman and inventor. In 1662 he organized the earliest system of hansom cabs in Paris[58] and it is said that, probably in 1658, he designed the well mechanism that can still be seen in the farmyard at the back of the house, Les Granges, where the Solitaries lived and taught. He is also thought to have invented the wristwatch: according to his niece Marguerite Périer (**X 318 n. 1**),[59] he wore his watch on his left wrist.[60] His inventiveness even extended to educational methods.[61] Not only did he draw up the so-called 'Elementary Geometry' and help to draw up a *Port-Royal Grammar*, he also coached his seven-year-old nephew Louis Périer in logic and mathematics (1378);[62] he was even found giving seven or eight street urchins a reading lesson (1454) – evidently sharing with them the benefit of his new reading method. Likewise, his *Discourses on the Worldly Condition of the Great*, a three-part meditation on the duties and responsibilities imposed by high worldly rank, were intended for the use of another boy, although one of very much higher social standing.

Not long after he had divested himself of all the superfluities of life, he even welcomed a pauper family into his own home in the Rue des Francs-Bourgeois Saint-Michel (now 54 rue Monsieur-le-Prince), sharing his simple meals with them. When one of their children was stricken with smallpox, he moved in, on 29 June 1662, with his sister and brother-in-law Gilberte and Florin Périer at their rented house (now 67 Rue du Cardinal-Lemoine) in the Faubourg Saint-Marcel. Early in July 1662 six weeks (42, 47) of acute illness set in, during which he was often visited by Fr Beurrier, the parish priest of Saint-Étienne du Mont.[63] He died a very painful death on 19 August 1662, having endured twenty years of physical distress; his discomforts and disabilities had been particularly severe during the last two years of his life (47). 'May God never abandon me' – echoing a phrase from his own *Memorial* (913*) – are said (34) to have been his dying words. The likeliest cause of his death was carcinomatous meningitis consequent upon a primary carcinoma of the intestines. But theories as to the cause of his death abound, and it is also possible that he suffered from polykystosis in his final years and that death may have come as the result of one or more aneurisms at the base of his skull.[64] He was buried on 21 August 1662 at the church of Saint-Étienne du Mont in Paris.

Pascal never married, though rumour had it that in 1652 he was involved in a romance, probably Platonic (37).[65] However, the lineage of his family is by no means extinct. Many descendants of cousins of his live today in France, Italy and England.

At least eleven death masks of Pascal are known to exist.[66] They are based on the original which in Louis XVI's reign belonged to the medallist Benjamin Duvivier. This subsequently became the property of Augustin Gazier[67] and now belongs to the Société de Port-Royal. In 1880 Gazier allowed the cast to be made which is now on display at the Musée du Ranquet, Clermont-Ferrand. Also cast in plaster of Paris from Duvivier's original were the death masks now at the Bibliothèque Sainte-Geneviève, Paris; the Musée National des Granges de Port-Royal; the Musée Carnavalet, Paris; the École de Médecine, Paris; the Musée Condé, Chantilly; Rouen; Newnham College, Cambridge,[68] and those in private hands.[69] It is said that another mask is still reverently preserved in a secret place in the church or presbytery of Saint-Étienne du Mont.

Nine portraits of Pascal date from the seventeenth century with varying degrees of probability (1361).[70] His friend Jean Domat's red-chalk sketch of him as a young man is the only one which could have been executed during his lifetime; it probably dates from about 1640.[71] A portrait in oils on canvas by François Quesnel the younger seems to date from 1664;[72] as Quesnel had never met or even seen Pascal, he worked from the death mask. During the last decade of the seventeenth century Gérard Édelinck produced two engravings of Quesnel's work, in the first of which (produced in or about 1691) Pascal looks to the left[73] whereas in the second (dating from 1697) he faces right; a portrait based upon the earlier of these engravings, and probably painted about 1700, used to belong to Prosper Faugère and is now at the Musée de Versailles.[74] The portraits at Amersfoort in the Netherlands, in the former palace of the archbishops of Toulouse, and at the Musée National des Granges de Port-Royal are based either on the Quesnel original or on one of the Édelinck engravings.[75] The painting, traditionally of him, showing the Périers' Château de Bienassis in the background is thought to have been produced towards the end of the seventeenth century, perhaps about 1690: the face may be by Quesnel.

Eight specimens of Pascal's mechanical calculator are known to have survived out of the probable total of between twelve and thirty manufactured. Four, including the one dedicated by its inventor to Pierre Séguier, are at the Musée du Conservatoire National des Arts et Métiers in Paris. Another is at the Musée du Ranquet, Clermont-Ferrand, another at the Staatlicher Mathematisch-Physikaler Salon,

Dresden,[76] another belongs to the computer company IBM. Most are like the IBM model in that they have eight wheels and are intended for calculations in six figures plus *sous* and *deniers*. The Dresden model is, however, the most elaborate in that it is equipped with ten wheels, whilst the specimen in private hands has five wheels for the purposes of addition and subtraction together with three that are calibrated in feet, inches[77] and *lignes* (*lignes* being twelfths of a French inch in the mid-seventeenth century).[78]

Pascal was a man of slight build with a loud voice[79] and somewhat overbearing manner (**XII ccliii**).[80] Like Nietzsche two or so centuries after him, he lived most of his adult life in great pain. He had always been in delicate health, suffering even in his youth from migraine and dyspepsia; from the age of eighteen onwards he never spent a day without pain (7). By the age of twenty-four he was an incurably sick man (10) and may even have experienced motor neuropathy about this time;[81] hence, perhaps, his statement that 'sickness is the proper state for a Christian' (32). From the age of thirty-five onwards he was in a permanent state of pain and debility, believing that he would die earlier than he in fact did (20). His health deteriorated greatly during the last four years of his life (12–13, 19), when he suffered from insomnia, persistent bowel disorders and from a form of paraplegia which may have been hysterical in origin and which was perhaps also complicated by intestinal and peritoneal tuberculosis. He could digest only asses' milk[82] and in August 1660 could no longer ride a horse and was hardly even able to walk (522).[83] He may also, towards the end of his life, have suffered a strange kind of vertigo, imagining that there was a yawning gulf beside his chair, from which he could only be protected by another chair, and that he was thus once again – as on Neuilly bridge – precariously poised on the very edge of an abyss (1365).[84]

As for his attributes of mind and character,[85] he was a man of formidable intellect[86] and prodigiously retentive memory,[87] versatile (in his contributions both to the mathematical sciences and to literature), sturdily self-reliant (being perhaps the greatest of all self-taught people), precocious (writing his *Essay on Conic Sections* at the age of sixteen), stubbornly persevering (in his construction of the mechanical calculator), a perfectionist, pugnacious to the point of bullying ruthlessness (in his controversies with Forton, Noël, La Loubère and even his sister Jacqueline) yet seeking to be meek and humble (perhaps immoderately so during his final years: 47), lucid (in his mathematical demonstrations), ludic (in his fondness for disguises), restlessly energetic (in his composition of the *Provincial Letters*), ambitious (in his project for the defence of Christianity), courageous (in his fight against illness), self-sacrificing (in his love of the poor), impatient (in the Wager argument) and perhaps also slightly unscrupulous (in his conduct of the cycloid competition). First with his *Essay on Conic Sections*, next with his mechanical calculator, then with his experiments into atmospheric pressure and finally with his mathematical work (not to overlook the religious writings), he worked not only in flashes of genius but in fits and starts; and this, perhaps, was as much the result of an attitude of mind as it was the direct consequence of his ill-health. That attitude of mind was essentially practical: he tackled a problem when there was a need for him, or for someone, to do so. Moreover, he felt increasing doubts about the value of any mathematical or scientific work. As these doubts grew, so too did the intensity of his religious beliefs. His personality was not only intense but also somewhat abrasive; he was always eager to outmatch others. Yet, at the same time, he had an instinct for self-abasement, believing that no gentleman should use the words *I* or *me*: 'he was in the habit of saying that Christian piety annihilates human *egoism*, whilst human civility conceals and suppresses it'.[88]

Being so dedicated to causes, Pascal was the very reverse of a voluptuous man. His sole instinct for beauty seems to have lain in the matchless perfection of his command of French prose: this prose, however, was hastily written, often in the fragmentary manner of the *Thoughts*: it is somewhat astonishing, therefore, that he wrote in Letter XVI of the *Provincial Letters*: 'I have only made this one longer because I have not had the time to make it shorter' (865; 257); such was his instinct for literary perfection.[89] Pascal was a misogynist, who would not allow any woman's beauty to be mentioned in his presence (23). According to his sister Gilberte, he even disliked seeing her caress her own children (24); marriage, he wrote to her, 'is a sort of homicide or, as it were, deicide' (521). It may well be that he was a victim of neurosis and that the 'night of fire' was the product of a hysterical temperament. He was, finally, a wonderful speaker whose greatness showed through, in the very power of speech, to his listeners: as when, *circa* May 1658, he addressed the Solitaries of Port-Royal concerning his plan for a defence of the Christian religion (149*),[90] or in his discussion with Isaac de Saci which was one of his earliest recorded utterances on religious matters.

Like Wren, Fermat and Leibniz, Pascal published little during his own lifetime – and very little indeed to which his name was attached. His fondness for disguises meant that at the time of his death in 1662 he was the self-acknowledged author of the *Essay on Conic Sections, New Experiments Concerning Vacuums,* and *Narrative Account of the Great Experiment on the Equilibrium of Liquids* (all early works, and the last of these largely consisting, in any case, of a letter from Florin Périer).[91] Moreover, it did not seem to the world at large that Pascal had any mathematical achievements to his credit other than the little known *Essay on Conic Sections.* The authorship of the *Letters to Carcavi* was clear, however, to Huygens's correspondent Ismaël Boulliau[92] – and probably also to most of Pascal's intimates, who were definitely aware of their friend's momentous correspondence with Fermat on probability theory.

The gradual but unsystematic publication of Pascal's work from 1663 onwards had much to do with the growth of his fame. Writing a fortnight or so after his death, Pierre Nicole feared that the high reputation merited by his brilliance of mind and personality, and so well appreciated by the small circle of his personal admirers, might die with him (50).[93] But by the turn of the century he had become notable enough to join St Vincent de Paul, Richelieu, Mersenne, Gassendi, Poussin, Descartes, Corneille, Boulliau, Turenne, Colbert, Molière, Racine and others in Charles Perrault's two-volume record of outstanding seventeenth-century Frenchmen – and actually to appear in the first volume.[94] Voltaire felt particularly exasperated by his insistent awareness of an eternal dimension to human life: in the last of his *Philosophical Letters* he sets out to demolish the *Thoughts,* intending 'to tear Pascal's skin without making Christianity bleed';[95] nevertheless, the epistolary format of the *Provincial Letters* may well have influenced Voltaire's choice of the same medium when writing about England. There was nothing, however, in Pascal's outlook and message which could greatly please the *philosophes* of the eighteenth century, with their thoroughgoing rationalism, their deism and (as time moved on) their atheism.[96] D'Alembert, on the other hand, appreciated his mathematical work. But to the agnostic libertarian Condorcet his *Thoughts* seemed to be nothing less than 'an indictment of the human species'.[97] Not until the very early nineteenth century was he 'rediscovered',[98] this rediscovery having been prompted by Charles Bossut's more comprehensive publication of Pascal's works in 1779. He appealed to the Romantics as polymath and as the creator of an exceedingly fragmentary body of work whose crowning achievement, the *Thoughts*, was doubly enigmatic in that it was unfinished. He was seen as a 'poet' in the Romantic sense of that word: both scientist and literary man, creative in his research, not only a discoverer but an inventor, and a seeker of infinite truths.

These same polymathic qualities, slightly differently viewed, made him appear to the later nineteenth-century reader and critic[99] as a *savant* of outstanding merit in the sciences and the humanities alike; he thus came to be regarded, by and large, as a man exemplary in his private life and little short of miraculous in his dedication to so many branches of learning during so brief an existence. He seemed to many, in that age of *Essays and Reviews*, the Tübingen Higher Criticism and Renan's *Life of Jesus*, to be pure and exalted in his devoutness – and no longer misanthropic or, worse still, superstitious as so many eighteenth-century commentators (Condorcet, for example)[100] had considered him to be.

Throughout the twentieth century[101] opinions about him have been sharply divided. By most, though not all, Catholic theologians and social thinkers he has been keenly admired. To non-Catholics he has often appeared bigoted and divisive, *dévot* rather than devout, the embodiment of an archaic piety and an intransigent religious formalism.

Pascal has been variously described as 'that remarkable, or rather matchless, intellect' (Gassendi, 1648),[102] a man 'more subtle than solid' (Nicole, *circa* 1672),[103] 'one of the most sublime minds the world has produced' (Bayle, 1696)[104] yet 'blindfolded' in the matter of cycloids (Leibniz, 1703),[105] 'a sublime misanthropist' (Voltaire, 1734)[106] yet one of 'the most enlightened men the world has seen' (Vauvenargues, 1746),[107] a 'saint' who believed in the 'most ridiculous superstitions' (David Hume, 1750),[108] 'a proud arrogant man feigning humility' (Chénier, *circa* 1790)[109] and a 'terrifying genius' (Chateaubriand, 1802),[110] who 'exerted a decidedly harmful influence upon French literature' (Friedrich von Schlegel, 1812)[111] yet was the author of 'incomparable letters' (Macaulay, 1832)[112] and 'the greatest of Frenchmen' (Disraeli, 1844),[113] a 'Christian hero' (Sainte-Beuve, 1848)[114] and a man important enough to have a day named after him in the Positivist Calendar (Auguste Comte, 1849)[115] yet who was 'an absolute shit' (Théophile Gautier, 1863),[116] a writer of incomparable prose (Renan, 1863)[117] who turned everyday things into the greatest ones (George Eliot, 1878),[118] a man 'alive to all greatness and all beauty' (Dean Church, 1875)[119] than whom there is hardly anyone 'who more truly bore the likeness of Christ' (Benjamin Jowett, 1881),[120] 'the greatest of all Christians' (Nietzsche, 1881)[121] yet the very epitome of the manner in which Christianity can corrupt human nature (Nietzsche, 1887),[122] 'the only *logical* Christian' (Nietzsche, 1888),[123] a neurotic (Daudet, 1889)[124] and 'a soul permanently ill at ease' (Walter Pater, 1895),[125] 'another Frenchman of pessimistic natural temperament' (William James, 1902),[126] 'a heaven-born mathematician' (Sir Leslie Stephen, 1898),[127] 'an extremely pernicious influence' (Claudel, 1911),[128] a 'sheer genius' and one of 'the very greatest writers who have lived upon this earth' (Lytton Strachey, 1912),[129] the 'sublimest of our intellectual leaders' (Barrès, 1918)[130] and the 'most powerful writer in the French language' (Paul Bourget, 1922),[131] 'mad' in his liking for 'gulfs and horrors' yet raising 'all the major problems of philosophy and conduct' (Aldous Huxley, 1929),[132] a 'domineering genius' (Jean-Raoul Carré, 1935)[133] who was 'a hard-hearted skinflint . . . bristling with hatred' (Charles Maurras)[134] but also 'a great soul grappling with all the pain of life' (Duhamel, 1941),[135] a sufferer from 'morbid mental tortures' (Bertrand Russell, 1945),[136] '[bankrupt] of historical method' (R.G. Collingwood, 1946),[137] who is 'religion itself' (Julien Green, 1949)[138] yet 'startlingly reminiscent of the Pharisee' (Ronald Knox, 1950),[139] a man 'imperishably youthful' (Jacques Chevalier, 1953, vii) yet 'still belong[ing], spiritually speaking, to the pre-Reformation era' (F.T.H. Fletcher, 1954),[140] 'the most important precursor of modern existentialism' (David Roberts, 1957),[141] 'one of the most penetrating minds of all time' (Christopher Hollis, 1968),[142] an 'antimetaphysical' thinker (Don Cupitt, 1980)[143] – one of the askers of great questions.

In Malcolm Hay's words,[144] 'few great men of letters who were so unproductive as he was have been so generously praised by posterity'. Nor has any writer except Goethe (the sage of Weimar) been so transformed after his death. He has, in short, been raised to the mythological status of a cult figure, or scapegoat, symbolic of the values of a particular civilization or way or view of life. And sometimes his characteristics have been generalized to the point of seeming representative of human nature as a whole. His symbolic status – something quite independent of the man, and the thinker, as he 'really' was – is particularly evident in the perceptions of him expressed by Voltaire, Nietzsche and Aldous Huxley; these perceptions were generally unfavourable.

But hardly anybody in the history of the world has been so much pondered and written about, at any rate by intellectuals. He is almost unique amongst the great literary artists in that, filled with his personal vision of things, he wishes to convince his fellow men of its universal and objectively demonstrable truth.

Pascal and his era occupy an extraordinary position in the spiritual and intellectual history of Europe. In so far as moral precepts of life are concerned, his view of Christianity is little different in its austerity from Milton's. Like so many of his lesser contemporaries, he lived within a devout Christian framework, and within a devout Christian community which set great store by purity and integrity of conduct. For him and so many of his scientific contemporaries, even including Newton, religion was a vital issue; but in his life it became 'all in all'.[145] Never again has an interest in mathematics and an interest in the religion of the Gospel been combined in one person to such a high degree.

2
Foundations

In mathematics Pascal was a child prodigy. By the age of fourteen or fifteen, having mastered Desargues's bafflingly abstruse work on synthetic projective geometry, he had become a fervent admirer of the *Brouillon Projet*. In February 1640, when he was sixteen years old, he published his own *Essay on Conic Sections*.[1] Starting with the intersections of circles, he seems (in accordance with the methods of Desargues's projective geometry) to have extended his proof of the Mystic Hexagon to ellipses (or *antobolas*), parabolas and hyperbolas. The Mystic Hexagon, discovered but unpublished, soon became the key to his understanding of this area of mathematics. From the two lemmas of the *Essay on Conic Sections* he is said to have deduced over 400 corollaries (58), including most of the propositions contained in the *Conics* of Apollonius of Perga. He believed, therefore, that the analytical co-ordinate geometry of which Descartes gave his first published exposition in 1637[2] was by no means the only way of elaborating a complete theory of conics: the same could be done by means of the Mystic Hexagon and the projective methods which Desargues had pioneered.[3] But he never achieved this unified theory and it was left to Poncelet and Chasles to carry the study of conics to its fullest development, thereby producing a general theory of transformations.

The mechanical calculator devised by Pascal in 1642 and brought to production standard by 1645 was one of the earliest in the history of arithmetical computing;[4] it was the direct development of the abacus.[5] In its method of operation[6] it was made more complicated by the circumstances of the French currency. Étienne Pascal did not compute his tax assessments in tens but by a method identical to that which prevailed in Britain until as late as 1971: in 1642, and indeed until 1799, a *livre* consisted of 20 *sols* and each *sol* was equivalent to 12 *deniers*. Of these two factors (the Babylonian dozen, as a fifth of sixty, and the Celtic score) by far the more awkward technical problems were posed by the duodecimal system. The fact that the *livre* had a unit base of 240, not 100, was largely responsible for the many delays which Pascal experienced in bringing his

Figure 2.1 A Theorem Inspired by Gérard Desargues

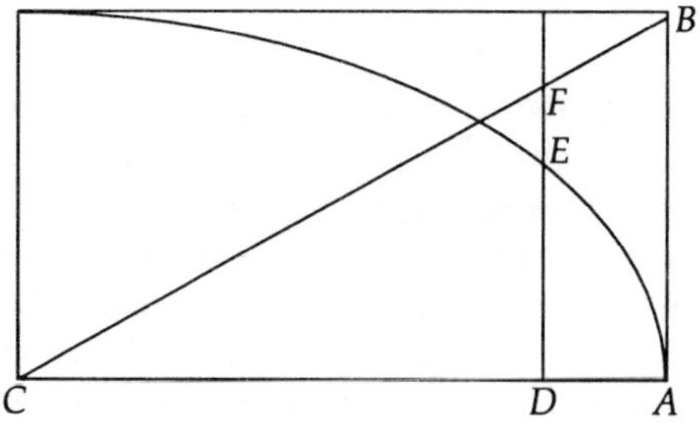

Given an ellipse with C at its centre, A at an extremity of its major axis, AB a line touching this extremity, and B the summit of the rectangle circumscribing its curve.

Let E be a point on the conic, with a line parallel to AB passing through E and intersecting CA at D and CB at F.

It follows that

$$\mathrm{DE}^2 + \mathrm{DF}^2 = \mathrm{AB}^2$$

This is the equation of the ellipse, since

$$\frac{\mathrm{DF}}{\mathrm{AB}} = \frac{\mathrm{DC}}{\mathrm{CA}}$$

This being so,

$$\mathrm{DF} = x\frac{b}{a}; \quad \text{therefore DE} = y; \quad \text{hence } y^2 + x^2\frac{b^2}{a^2} = b^2$$

$$\text{and therefore} \quad \frac{x^2}{a^2} + \frac{y^2}{b^2} = 1.$$

In 1639 this theorem was foreshadowed in Desargues's *Brouillon Projet*. More clearly restated in the Essay on Conic Sections in the following year, it shows the extent of Pascal's indebtedness to Desargues, who was the co-founder, with him, of projective geometry.

machine to perfection, and for most of his technical anxieties. Otherwise, the substance of the invention was simplicity itself.

Side by side in an oblong box were placed six small drums, round the upper and lower halves of which the numbers 0 to 9 were written, in descending and ascending orders respectively. According to whichever arithmetical process was currently in use, one half of each drum was shut off from outside view by a sliding metal bar: the upper row of figures was for subtraction, the lower for addition. Below each drum was a wheel consisting of ten (or twenty or twelve) movable spokes inside a fixed rim numbered in ten (or more) equal sections from 0 to 9 etc, rather like a clockface. Wheels and rims were all visible on the box lid, and indeed the numbers either to be added or subtracted were fed into the machine by means of the wheels: 4, for instance, being recorded by using a small pin to turn the spoke opposite division 4 as far as a catch positioned close to the outer edge of the box. The procedure for the basic arithmetical process, addition, was then as follows.

To add 315 + 172, first 315 was recorded on the three (out of six) drums closest to the right-hand side: 5 would appear in the sighting aperture to the extreme right, 1 next to it, and 3 next to that again. To increase by one the number showing in any aperture, it was necessary to turn the appropriate drum forward $\frac{1}{10}$th of a revolution. Thus, in this sum, the drum on the extreme right of the machine would be given two turns, the drum immediately to its left would be moved on $\frac{7}{10}$ths of a revolution, whilst the drum to its immediate left would be rotated forward by $\frac{1}{10}$th. The total of 487 could then be read off in the appropriate slots. But, easy as this operation was, a problem clearly arose when the numbers to be added together involved totals needing to be carried forward: say, 315 + 186. At the period at which Pascal was working, and because there had been no previous attempt at a calculating-machine capable of carrying column totals forward, this presented a serious technical challenge.

Linking the ten movable spokes to their appropriate drum was an elementary gearing mechanism of five toothed wheels. Pascal overcame the problem of transfers to the power of ten not without considerable hardship but in a way remarkable, in the event, for its operational facility. He invented a *sautoir* (or escapement arm) whereby a pawl attached to a counterweight regulated the motion of a wheel representing a power of ten, through contact with a tooth of that wheel. Whenever, for example, the unit wheel had completed a revolution, the counterweight would rise automatically whilst the pawl would slide away to the right, working loose from the tooth of the decadic wheel against which a spring normally held it in place. Still in contact with the wheel, however, the pawl would push it on leftwards by $\frac{1}{10}$ of a revolution as the counterweight fell back.

Subtraction was also possible on Pascal's calculating-machine. This was not as straightforward an operation as it might seem, because the wheels and gearing mechanism could not be put into reverse. Consequently, Pascal devised a system of parallel notations enabling the converse arithmetical process to be carried out with the minimum of complexity. The drums that were used for addition were also numbered in reverse from 9 to 0, and the only further and inevitable refinement was occasioned by the problem of negative transfers.

Pascal does not seem to have been aware of Wilhelm Schickard's so-called calculating-clock, manufactured as early as 1624.[7] This machine apparently had the advantage over Pascal's in that it could easily perform the non-linear operations of multiplication and division,[8] whereas Pascal's showed great awkwardness in those operations despite its inventor's claim (**I 300**) that it could perform all four of the processes of arithmetic (the Dresden model is, in fact, capable of the four). Nine years after Pascal's death Leibniz invented a more versatile machine capable of the four arithmetical processes, which, thanks to a gearwheel or (more precisely) a *stepped* wheel, performed the operations of multiplication or division by means of repeated additions or subtractions.

Yet not even Leibniz's desktop calculator was particularly reliable.[9] An exhibition held at the Science Museum, London in 1992 has rightly called Schickard's, Pascal's and Leibniz's machines 'ornate curiosities': 'for all their ingeniousness', a caption read, 'these devices were . . . serious-minded playthings that did not always work reliably and were unequal to the demands of routine use'. Great progress was made during the next 150 years. Müller's calculator could not only perform the four arithmetical processes but also calculate to fourteen figures. Making use of the method of finite differences (whereby second-order polynomials have the property that the second difference is constant), Babbage invented a 'difference engine'[10] in 1822 which produced arithmetical values by repeated addition: its true forerunner in this respect was Leibniz's, not Pascal's machine. The latter directly foreshadows the cash register, but the mechanical calculators invented by Leibniz and Babbage have been the really formative influences upon the electronic calculating-machine and the analogue and digital computers.[11]

Nevertheless, Pascal's desktop mechanical calculator was a considerable triumph for so young a man. He was, after all, only nineteen when he first began work on it and twenty-two years of age when he brought it to rough-and-ready completion; for this reason perhaps, more than for any relevance it may have to modern algorithms of computing, his name was given (by Niklaus Wirth) to one of the earliest of computing languages in 1971. Most characteristic of its inventor was the way in which Pascal's calculator was designed to fulfil a particular need. Though Napier had published his system of logarithms in 1614, this ingenious technique both for multiplying and dividing and for involution and evolution was not of any real value to Étienne Pascal. There is evidence that Blaise intended to go on from the desktop calculator to a machine capable of extracting square roots; but such a device would have had no immediate utility, and it is eminently characteristic of Pascal that he appears to have taken it no further. The second phase of his scientific activity ends on a practical note.

It was, however, in the statics of fluids that Pascal's contributions to physical theory were made. In experiments at Rouen between October 1646 and October 1647, using columns of mercury, water and wine, some of them fastened to ships' masts, he or his friends measured the heights to which these fluids would rise when (as we now know) they are counterbalanced by the external atmosphere. He thus discovered that there is some force which resists the formation of a void either real or apparent – whatever it is that is produced at the upper end of the experimental tube – and that, within narrow limits at least, this force is measurable and constant. It is measured, for example, by the tendency of a column of mercury 27" high to run down, being equal, in a tube of uniform diameter, to the weight of such a column. Any force greater than this will produce a void. If, therefore, a column of some other substance is used, the height of that column must be to the height of mercury inversely as the density of the first is to the density of the second.

This point is established in *New Experiments Concerning Vacuums* (362–70), the pamphlet in which he reported some of his recent scientific research (8 October 1647).

In this pamphlet, however, Pascal would only venture the personal opinion that whatever was produced at the upper end of the experimental tube was a vacuum; he would not even rule out

Figure 2.2 Pascal's Historic Vacuum Experiment

Take a tube which is curved at its bottom end, sealed at its top end A and open at its extremity B. Another tube, a completely straight one open at both extremities M and N, is joined into the curved end of the first tube by its extremity M. Seal B, the opening of the curved end of the first tube, either with your finger or in some other manner (with, for example, a hog's bladder) and turn the whole of the apparatus upside down so that, in other words, the two tubes really only consist of one tube, being interconnected. Fill this tube with quicksilver and turn it the right way up again so that A is at the top; then place the end N in a dishful of quicksilver. The whole of the quicksilver in the upper tube will fall down, with the result that it will all recede into the curve unless by any chance part of it also flows through the aperture M into the tube below. But the quicksilver in the lower tube will only partially subside as part of it will also remain suspended at a height of 26"–27" according to the place and the weather conditions in which the experiment is being carried out.

Now the reason for this difference is that the air weighs down upon the quicksilver in the dish beneath the lower tube, and thus the quicksilver which is inside that tube is held suspended in balance.

But it does not weigh down upon the quicksilver at the curved end of the upper tube, for the finger or bladder sealing this prevents any access to it, so that, as no air is pressing down at this point, the quicksilver in the upper tube drops freely because there is nothing to hold it up or to resist its fall (450).

the notion of Nature's *limited* abhorrence of vacuums (368), being unwilling to go beyond the provisional bounds of his knowledge in the matter. In October 1647 he had not yet proved to his own complete satisfaction that the apparent void is, or is not, a void.

Descartes, on the other hand, differed from Pascal, Roberval and, to some extent, Gassendi,[12] in his conviction – expressed in *The Principles of Philosophy*[13] – that a vacuum definitely could not exist within nature: he held, with Aristotle, that to assert the existence of vacuums was to be guilty of a profound contradiction, since the essence of matter was extension; the physical universe must be a closely packed plenum, containing no empty spaces.[14] Cartesian natural philosophy demanded *a priori* that as the air was driven out of the experimental tube by mercury, so it would rise to the upper limits of the atmosphere – displacing in its turn a quantity of the primary matter or ether which, descending to earth, would fill the space previously occupied by air; these cosmic displacements of matter, producing motion by contact as they swirled about in the ether, came to be known as the Cartesian vortices.[15] And so it was that when Pascal received two visits from Descartes, on 23 and 24 September 1647, the two men engaged in fierce argument concerning the existence of vacuums. To Descartes's assertion that 'rarefied matter' – or ether – was in the tube Pascal, says Jacqueline **(II 43)**,[16] 'replied as best he could'.

A month after the visits from Descartes, the publication of *New Experiments Concerning Vacuums* led Pascal almost at once into further controversy, this time with the Jesuit and neo-Aristotelian priest Étienne Noël (October-November 1647;[17] March or April 1648).[18] The latter was not an intellectually fastidious man. Behind him he had a long and distinguished career in the Church. He was a Jesuit of eminence, who had taught Descartes philosophy (though the pupil had rapidly excelled the master) and who had been Rector of the colleges of Eu, La Flèche, and the Collège de Clermont in Paris. The whole emphasis of his writings was to harmonize the teachings of the medieval Church and the Ancients, especially Aristotle, with the findings of modern thought: but he was much more concerned to underline those aspects of traditional thought with which modern thinkers agreed than to correct the ancient teachings in the light of newly established truth. Noël was a scholastic, with a profound and indeed humble respect for the Church's perennial wisdom: no malicious or hostile man, his sole purpose in writing to Pascal, within a few days of reading *New Experiments*, was to convince himself as much as his correspondent that, despite all these strange and disconcerting developments in physics, the age-old belief in the non-existence of vacuums could still be maintained. For God to have produced a vacuum would, he believed, mean that He had willed the destruction of the world. Such was the kindly priest with whom Pascal now came into conflict. The conflict was of the priest's own choosing; it was Noël's decision to write the first letter (1438–42), his wish that some provisional accommodation between the old and the new learning might be possible. In this perhaps laudable but certainly foolish attempt he was the first Jesuit to smart under the thrust of Pascal's invective and satire.

Noël, as a matter of fact, was even ready – in his own mind, if not in his public teaching – to abandon his notion that a form of rarefied air might have entered the experimental tube through the pores of the glass. Privately at any rate, he accepted the notion of atmospheric pressure; indeed, at this time (November 1647) he was perhaps even more convinced of it than was Pascal himself. But though he would go so far with Pascal, he was not prepared to acknowledge the existence of vacuums which to him was an ontological absurdity. Pascal, on the other hand, defended both the theory of atmospheric pressure and the existence of the void. His only subsequent comment on the Noël controversy came in an important letter to his friend François Le Pailleur in February 1648: Noël, he complained, had failed to distinguish between the intellectual act of defining a vacuum and the scientific assertion that such a thing exists (377–91).

It was on 15 November 1647, at the height of the controversy over *New Experiments Concerning Vacuums*, that Pascal wrote (392–5; **II 153–62**) to his brother-in-law Florin Périer, at Clermont-Ferrand, to ask him to prepare the definitive experiment – for Florin, like his wife's father and brother, was keenly interested in the natural sciences, and from where he lived a mountain 1465 metres high, the Puy de Dôme, was within easy reach. The aim of this experiment was to establish beyond any doubt that the vacuums which the Rouen experiments had shown to be theoretically possible did indeed exist, that there was not even a *limited abhorrence* of vacuums within Nature and that all this could be demonstrated by means of experiments at different altitudes. What with bad weather, delays in procuring the necessary apparatus, and Périer's own duties as an assessor at the Court of Customs and Excise at Clermont, these experiments were delayed, however, until 19 September 1648.

But prior to the Puy-de-Dôme experiment another decisive step had been taken in the history of physics. This was the creation of self-enclosed vacuums: an experiment which, in his letter to his brother-in-law (393–4; **II 155–8**), Pascal claims that they had already conducted very early in November 1647 but one which is equally attributable to Pascal's companion in the Forton affair, Adrien Auzout. Certainly it seems that Auzout, about June 1648, performed the most convincing demonstration of a vacuum within a vacuum, incidentally demonstrating that both these young men had the most substantial intellectual achievements to their credit and were no mere bigoted extirpators of metaphysical heresy. As carried out by Auzout, this experiment consisted of a long test tube opening out at its upper end into a large glass balloon sealed by an impermeable membrane: within the balloon a small trough of mercury was placed and a pipette, open at both ends, suspended in it without the lower end of the pipette touching the bottom of the trough. The whole of the double apparatus was filled with mercury; the upper end of the pipette was sealed; the lower end of the long tube was immersed in a larger trough of mercury; the long tube was then unsealed at its lower end; within the long tube the mercury then fell to its usual level; but within the pipette the mercury completely drained out into the smaller trough. Thus it was established that, with no air pressure within the glass balloon to sustain the mercury in the pipette, the column of mercury would not hold up at all. Conversely, when the impermeable membrane was pierced at the top of the whole apparatus, air rushed in, the mercury within the pipette rose under the pressure of the atmosphere bearing down upon the higher trough, whilst the mercury in the long tube drained out into the trough below. This, whichever of the two young men devised it, was an experiment which definitively refuted the Cartesian hypotheses of vortices and ether.

Working in mathematics as well as in physics, Pascal had to wait until September 1648 for the final and decisive step to be taken in his contribution to physics. Why he did not himself travel to Auvergne to be near, if not present at, his great experiment is not known. The reason for his absence from the scene was probably medical. As late as 1780, it took eight days to travel by coach from Paris to Clermont. But he was informed of the complete success of the experiment in a letter written by his brother-in-law three days after its occurrence (395–9).

Early in the morning of 19 September 1648 Florin Périer had assembled a numerous company of learned men and functionaries of the city, including the superior of the Minim Friars, a canon of Clermont Cathedral, a doctor, a relative of Pascal's on his mother's side, and two of Périer's own colleagues from the Court of Customs and Excise. In the convent garden, which has now been absorbed into the Place de Jaude, they performed their experiment for the first time. In two identical test tubes they measured the fall of the mercury; in both the level of mercury stabilized at $26\frac{3}{10}$ (French) inches (of 1648) above the level of mercury in the trough. A Minim, Fr Chastin, was asked to make periodic observations of the one test tube they were proposing to leave behind; with an excitement which can scarcely be realized today (to us the experiment seems simple) they then set off on the ten-mile ascent of the Puy de Dôme. Following the Roman road, some paving-stones of which still exist, they climbed to La Font-de-l'Arbre: the mercury stood at 25 inches! From there it was a further four steep miles to the top of the Puy: the mercury stood at $23\frac{1}{6}$ inches! In order to prove his brother-in-law's point beyond any contradiction, Périer performed the experiment five more times, at five different places on the summit of the mountain: once sheltering in the little chapel erected on the site of a Roman temple but now destroyed; then in the open air, in the sunshine, in the wind and in the misty drizzle. Each time the reading was constant, at $23\frac{1}{6}$ inches! For further corroboration of the theory, they halted again at La Font-de-l'Arbre on the way down: and again the reading at that altitude remained unvaried at 25 inches! From Fr Chastin they established, on their return to the convent garden, that the mercury level in his test tube had never varied from the original reading of $26\frac{3}{10}$ inches. Their final verification was to compare the mercury levels in both tubes, and to find that they were identical. Next day, to make assurance doubly sure, Périer performed the experiment for the eleventh time: on this occasion, at the very top of Clermont Cathedral where the reading was $26\frac{1}{10}$ inches.

No precision was lacking in the way Périer had set about this experiment: the numerous verifications, so numerous as to be almost superfluous; the making of these verifications in varying atmospheric conditions; the plentiful witnesses, all of them men of good repute and standing, some even with a passable knowledge of the natural sciences. Mosnier, in particular, was a man well able to grasp the striking originality and fundamental importance of his day's work. In a letter to Gassendi he provided an independent record of Pascal's – and Périer's – achievement; and Gassendi, in a letter (previously quoted) to his printer Barancy, has recorded his admiration for Pascal's experimental boldness. Viewed in a longer retrospect, the fundamental importance of perhaps the one great experiment in scientific history at which its originator was not present is threefold: it led directly to the invention of the barometer, and thus to the creation of meteorology (for years an important meteorological station stood on the summit of the Puy de Dôme); it led also to the first practical and scientific method of determining altitudes; and from the knowledge of the existence and properties of vacuums came the invention of the hydraulic press.

Was Descartes the true originator of the Puy-de-Dôme experiment, Pascal's most brilliant and lasting contribution to the realm of physics? Subsequently, he certainly claimed to have sown this idea in the young man's mind: 'I had advised M. Pascal', Descartes writes to Mersenne,[19] 'to experiment whether quicksilver rose as high when one is on the summit of a mountain as it does when one is down below; I do not know whether he has yet done so.' And again, enquiring from Carcavi whether the experiment had yet taken place: 'I would be entitled to expect to be told this by him rather than you, because it was I who advised him two years ago to conduct this experiment, assuring him that, though I had not conducted it myself, I was in no doubt of its success'.[20] And yet again, after hearing from Carcavi (not from Pascal) of the success of the experiment: 'It was I who asked him two years ago to conduct it, assuring him of its success – as would be entirely consistent with my own scientific principles'.[21]

Whether or not Descartes instilled the original notion of the experiment in Pascal's mind, one thing is certain: the experiment as Descartes would have conducted it would have served a different purpose. For whereas Pascal wished to demonstrate the existence of vacuums, Descartes, who (unlike Roberval and Mersenne) looked favourably upon the hypothesis of barometric pressure, would have sought to convert that hypothesis into a theory. Descartes would have performed the experiment in order to prove that at the top of a high mountain the mercury in the experimental tube would have risen less high because, the tube then being closer to the upper limits of the atmosphere, the displacements of air and ether would have been less great than at sea level.

Though Pascal could, or would, not be present on the Puy de Dôme, he himself conducted further verifications in Paris. The place he chose for his experiments was the tower of the church of Saint-Jacques de la Boucherie (399),[22] not far from his own parish church of Saint-Merri. This still exists as the Tour Saint-Jacques, one of the most notable of Parisian landmarks, though the main body of the church to which it was adjacent has long been destroyed. Pascal also tested his hypothesis on the top storey of a house, 'ninety steps high' (399), and (legend has it) on a tower of Notre-Dame cathedral. A little later a Notre-Dame experiment was unquestionably carried out by Jacques Rohault, who also verified Pascal's findings on the frozen surface of the river Seine.[23] Gassendi, on 5 February 1650,[24] pursued his own parallel investigations at sea level, halfway up a hill and on a hilltop outside Toulon. Other experiments were conducted at Dieppe (453, 463), the port nearest to Paris and therefore at sea level.

Pascal had not discovered the actual principle of the barometer but he had provided incontrovertible proof that vacuums existed. He had also demonstrated that the height of the column of mercury in a barometer decreases as it is carried upwards through the atmosphere. He soon (in November 1648) embodied his findings in his *Narrative Account of the Great Experiment on the Equilibrium of Liquids* (392–401),[25] thus marking his priority in this field of research and setting down results for other scientists either to build on or to contradict. He continued to refine his invention, working also on meteorology and no doubt systematizing his thoughts for the definitive 'Treatise on Vacuums' which he had been heralding ever since his account of the Rouen experiments in October 1647 but was never to bring to full fruition.

Pascal's most important single work on the statics of fluids is the *Treatise on the Equilibrum of Liquids*, composed about 1653: this brief but comprehensive treatise (412–28) formulates the well-known Pascal's Principle: that, expressed in modern terms, 'pressure applied to an enclosed fluid is transmitted equally and undiminished in all directions to all parts of the containing vessel, at right angles to its surface and with equal force upon equal areas'. This is the very basis of the science of hydrostatics: all subsequent studies on the stasis of liquids and gases are hugely in its debt. The *Treatise on the Equilibrium of Liquids* confirmed what earlier scientists such as Stevin and Mersenne had only dimly realized: that the weight of liquids varies according to their height. A second treatise, the *Treatise on the Weight of the Air Mass* (428–62), composed about 1654, dealt with the effects of the weight of the atmosphere largely in terms of the principles formulated in the first treatise: not entirely so, however, since the compressibility and elasticity of air were also taken into account. The *Treatise on the Weight of the Air Mass* computed the weight of the earth's atmosphere as 8,283,889,440, 000,000,000 lb (456);[26] it also led to the measurement of altitude from atmospheric pressure.

In these two *Treatises* Pascal foresaw some of the practical uses to which his discoveries could be applied: earth-digging operations (400), the lifting of huge weights (417), and weather-forecasting (467). Others have been the hydraulic press, hydraulic brakes, cranes, lifts, pumps, turbines and the syringe. Pascal's experiment of pushing mercury through capillary tubes even laid the foundations of modern anæsthetic procedures, which in 1846 involved the pushing of gases along tubes against varying resistances. These discoveries in physics entitle him to rank as one of the founders not only of hydrostatics but also of hydrodynamics.

Pascal returned to his scientific research in May 1653. For sixteen months until September 1654 he was greatly preoccupied by mathematics. His work on probability theory, combinatorial analysis and arithmetic gave rise to what was by far his most fertile scientific period; it was the only time when the scientific and social aspects of his life were in reasonably happy conjunction. In his 'Celeberrimæ Matheseos[27] Academiæ Parisiensi', or 'Address to the Most Celebrated Parisian Academies of the Mathematical Sciences'[28] (73–4, 1402–4), composed in (July?)[29] 1654, he outlined a wide range of future mathematical work, foreshadowing various treatises some of which have either vanished or else were not written.

Unmentioned in this 'Address' (except for the most glancing reference to the fact that they had been completed and were about to be published) are the *Treatise on the Equilibrium of Liquids* and *Treatise on the Weight of the Air Mass*. Both these works had indeed been completed yet, despite Pascal's assurances to the contrary, they remained unpublished until a year after his death. Also unmentioned in the 'Address' is a treatise on mechanics, since lost.[30] The work outlined to his colleagues is exclusively geometrical.

Of the research he highlights in his 'Address' (73–4, 1402–4) the following treatises have either disappeared or else were not written: 'Conical Contacts' (on the determination of conics from five given elements, points or tangents) and 'Solid Loci'[31] (which, by way of contrast to Descartes's analytical treatment of Pappus's problem,[32] used projective geometry to produce an alternative solution); 'The French Apollonius Improved Upon'[33] and 'Spherical Contacts', both of which were (or were intended to be) geometrical demonstrations of the contacts of spheres and circles;[34] three lesser geometrical exercises;[35] and 'On Magic Numbers', which would have dealt (or did deal) with that subject of perennial fascination to mathematicians, magic squares.[36]

Mathematical writings indicated to the Academies of the Mathematical Sciences, some major element of which remains, are 'The Complete Work on Conic Sections' and *De Aleæ Geometria* or *The Geometry of Chance* (75–171).

The first of these would have contained the so-called 'Treatise on Conic Sections', or (more accurately) 'The Generation of Conic Sections'[37] (66–70), which deals with conics generated from a circle by central projection. Greatly indebted to Desargues, it would have been the first part of Pascal's proposed treatise on conics, of which the youthful *Essay on Conic Sections* was merely the forerunner. Completed about 17 March 1648 but since lost, it is known to us only from the notes taken from it by Leibniz and Tschirnhaus.[38] It is the only one of the five distinct sections of the full-scale treatise which still exists in fairly complete form – indeed, probably the only one of the five to have come anywhere near a reasonable state of completion. Included in it is the so-called 'Pascal's Theorem' (58) on the Mystic Hexagon, or hexagon inscribed in a conic, which Leibniz has also preserved for posterity: this proves that the intersections of pairs of opposite sides of a hexagon inscribed within a conic section are collinear, demonstrating by means of projections that a conic is produced by the polar curve of any point on the hexagon. 'The Generation of Conic Sections' also includes further related work on the theory of poles and polars, on segments determined by a conic section, on secants parallel to a fixed direction and on tangents at a given point on a conic.

In his letter (63–5) to Étienne Perier returning, in August 1676, the precious mathematical manuscripts lent to him by Pascal's heirs, Leibniz urged (65) that the writings on conics should be published with all haste as another mathematician – probably Philippe de La Hire[39] – was known to be working on the same subject. But there seems to have been no sense of urgency in this race to a finishing-line which, quite simply, did not exist.

By far the most important part of the work outlined in the 'Address' is, however, *The Geometry of Chance*. This consists partly of the so-called *Règle des Partis*, or *Probability Calculus*, and partly of the *Treatise on the Arithmetical Triangle* (97–171).[40] The former is an exchange of letters, unpublished until 1679,[41] which Pascal conducted with Fermat. Of the five surviving letters (dated ?, 29 July, 24 and 29 August and 25 September 1654), two – the second and third in the extant series (77–89) – are from Pascal's pen.[42] Included with the *Treatise on the Arithmetical Triangle* were two further treatises on number theory, *Concerning the Recognition of Multiple Numbers by the Mere Sum of their Digits* and *The Summing of the Powers of Integers*.

In their five letters on the probability calculus Pascal and Fermat jointly founded a branch of mathematics which was almost entirely new.[43] Initiated in these letters and explored further in the work on the binomial expansion in the *Treatise on the Arithmetical Triangle*, this investigation of the probability calculus stemmed from Pascal's fascination with the games of chance which he now learned to play with his fashionable friends. The Chevalier de Méré was a keen gambler.[44] In his company and when with other fashionable friends such as the Duc de Roannez Pascal must have played or seen being played piquet, backgammon, reversi and various games of dice. The two questions to be resolved in the Fermat correspondence, both concerning gambling, were put to him by Méré (77). Within how many throws of two dice can I expect to come up at least once with a double six? (This is the so-called *dice problem*[45] already familiar to Cardano.) More importantly, how should the stakes be divided if a game of dice is interrupted when the two players have already scored an unequal number of points? (This is the so-called *problem of points*, or *division problem*,[46] which was known to the Italian mathematicians Pacioli, Tartaglia and Cardano as early as the fifteenth and sixteenth centuries.) But although Pascal and Fermat resolved the problem of points in respect of two players, the mathematics which they created – simplistic, according to J. M. Keynes[47] – was by no means capable of resolving it in respect of three. In order to ascertain the risks involved in a three-player game all 27 possible outcomes needed to be set out in tabular form.[48] Within this total of 27 outcomes the *règle des partis* (Fermat's and Pascal's formulation of the probability calculus in terms of the division problem) is actually incapable of computing the number of wins favourable to A, B or C.

Towards the end of Pascal's letter dated 24 August 1654 (letter iii) an error occurs in his mathematical reasoning, which is corrected by

Figure 2.3 Pascal's Triangle

Tartaglia's Triangle, after Niccolò Tartaglia (1500–57), is another name for this triangular array of binomial coefficients

	1	2	3	4	5	6	7	8	9	10	11	12
(1)												
(1)	1											
(1)	2	1										
(1)	3	3	1									
(1)	4	6	4	1								
(1)	5	10	10	5	1							
(1)	**6**	**15**	**20**	**15**	**6**	**1**						
(1)	7	21	35	35	21	7	1					
(1)	**8**	**28**	**56**	**70**	**56**	**28**	**8**	**1**				
(1)	9	36	84	126	126	84	36	9	1			
(1)	10	45	120	210	**252**	210	120	45	10	1		
(1)	11	55	165	330	462	462	330	165	55	11	1	
(1)	12	66	220	495	792	**924**	792	495	220	66	12	1

These are *figurate* numbers. Each integer in the Triangle is the sum of the one immediately above it and the one directly to the left of that.
The numbers in the second column down (1,3,6,10 . . .) are 'triangular' numbers, those in the third column (1,4,10,20 . . .) are 'pyramidal'.

The Triangle was discovered by the Chinese mathematician Chia Hsien in the twelfth century and popularized by Chu Shih-Chieh in 1303. It was converted by Pascal into a method of combinatorial analysis.

How many combinations are there of 10 things chosen 5 at a time? Pascal showed that this could be calculated by correlating the total column (vertical column 1) with the subset column (horizontal, underlined) of things chosen at one go.
Thus, the number of combinations of 10 things chosen 5 at a time is to be found at the intersection of the total column 10 with the subset column 5, i.e. 252 **(printed in bold type).**

Likewise, the number of combinations of 12 things chosen 6 at a time is to be found at the intersection of the total column 12 with the subset column 6, i.e. 924 (printed in bold type).

Pascal's Arithmetical Triangle also indicates the probabilities involved in throwing coins or dice. For example, the numbers in the sixth row from the top add up to 32, which is the number of ways in which five coins can fall. By putting each of the numbers in that row over 32 (i.e., $\frac{1}{32}, \frac{5}{32}, \frac{10}{32}, \frac{10}{32}, \frac{5}{32}, \frac{1}{32}$), it is possible to calculate the probabilities of the various different outcomes.
Pascal's Triangle possesses many other remarkable properties.
It gives the coefficients in the expansions of binomial expressions such as $(x+y)^6$.
(These rows of numbers are also printed in bold type.)
Such coefficients are for descending powers of x and for ascending powers of y going from left to right.

$$\text{Thus, } (x+y)^6 = (x+y)(x+y)(x+y)(x+y)(x+y)(x+y) = x^6 + 6x^5y + 15x^4y^2 + 20x^3y^3 + 15x^2y^4 + 6xy^5 + y^6.$$

$$\text{Likewise } (x+y)^8 = (x+y)(x+y)(x+y)(x+y)(x+y)(x+y)(x+y)(x+y) = x^8 + 8x^7y + 28x^6y^2 + 56x^5y^3 + 70x^4y^4 + 56x^3y^5 + 28x^2y^6 + 8xy^7 + y^8.$$

Fermat in his next[49] letter, dated 25 September 1654 (letter **v**). Here Fermat shows that if four games (with 81 possible outcomes), rather than three (with 27), are taken into consideration, the answer is nevertheless identical.[50]

It remained for Huygens[51] and Jakob Bernoulli, during the next fifty or so years, to develop the tools and skills which the probability calculus demanded. Rémond de Montmort and De Moivre supplied practical applications of the new methods in 1708 and in 1718 and 1725.

In his work on the binomial expansion in the *Treatise on the Arithmetical Triangle* (97–171)[52] Pascal investigated binomial coefficients and laid the foundations of the binomial theorem. This treatise provided the basis of the theory of combinations; it was the first to focus on that subject.[53] A triangular array of numbers consists of ones written on the vertical leg and on the hypotenuse of a right-angled isosceles triangle; each other element composing the triangle is the sum of the element directly above it and of the element above it and to the left. Pascal proceeded from this to demonstrate that the numbers in the $(n+1)^{st}$ row are the coefficients in the binomial expansion of $(x+y)^n$. Thanks to the clarity of his formulation of the problems involved, the Triangle, although very far from being original, is one of his finest mathematical achievements:[54] it has been wide-ranging in its influence upon future scientific development (including the theoretical basis of the computer). It has also made an essential contribution to the development of combinatorial analysis. Moreover, 'through the work of John Wallis it led Isaac Newton to the discovery of the binomial theorem for fractional and negative indices, and it was central to Leibniz's discovery of the calculus'.[55]

The full practical benefits of Pascal's pioneering work on probability were recognized at an early date.[56] In the realm of physics this pioneering work has helped to shape quantum theory, in which physical events can be known only in terms of probabilities; the techniques of combinatorial analysis evolved in the *Treatise on the Arithmetical Triangle* also enabled Clerk Maxwell and Boltzmann, working independently, to evolve the kinetic theory of gases. The same techniques underpin statistics[57] and the mathematical content of economics, and have given a powerful impetus to games theory and decision theory. The concept of *calculable risk* has been of enormous value in futures trading (despite the importance of the random-data element in the algorithm of its mathematical processes) and the same is also true, for example, of medical research. Moreover, combinatorial analysis has greatly influenced the technical input of actuarial calculations,[58] cybernetics, psephology, operational research and other aspects of industry,[59] commerce and politics in the twentieth century. The work undertaken by Pascal and Fermat in 1654 released the sciences from the straitjacket of absolute certainty, establishing instead the concept of a stochastic universe.

Concerning the Recognition of Multiple Numbers by the Mere Sum of their Digits (159–65, 1421–7) and *The Summing of the Powers of Integers* (166–71, 1427–32) – the appendices to the *Treatise on the Arithmetical Triangle* within the overall framework of *The Geometry of Chance* – have also had their seminal importance.[60] The first, admittedly more limited in scope than the second, studied the general arithmetical problem of determining whether any (non-zero) integer is a multiple of another and, if not, what remainder the divisor will produce; it is interesting as yet one further reflection of Pascal's fascination by arithmetic. In the second appendix (*The Summing of the Powers of Integers*),[61] however, he made a huge contribution to the theoretical problems of indivisibles, going far beyond Cavalieri's findings in *Six Geometrical Exercises*: thus he foreshadowed the development of the infinitesimal, and hence the integral, calculus. More particularly, he investigated the expression of the sum of the powers of like order of a sequence of numbers in arithmetical progression, arriving at the formula that 'the sum of the like powers of all integers is, to the power immediately above the last of those integers, as unity is to the exponent of that higher power' (171, 1432). Pascal's Recurrence Formula for the sum of the powers of integers was to become the standard textbook method.[62] In algebraic terms it may be expressed thus:-

$$(n+1)^r - (n+1)$$
$$= r\sum n^{r-1} + \binom{r}{2}^{r-2} + \binom{r}{3}\sum n^{r-3} + \ldots + r\sum n$$

where $\sum n^r$ is written for $\sum_{i=1}^{n} i^r$.

This was one of his most original contributions to mathematics.

Pascal also observed that the results of the summing of the powers of integers make possible the solution of certain quadrature problems. This was his first reference to Cavalieri's calculus of indivisibles, an area of mathematics which, although very recent, had already been fairly intensively studied.

So ended the fourth period of his scientific activity.

Physically, 1654 was a year of continuing ill-health: 'though I am still bedridden, I must tell you that yesterday evening I was given your letter', he wrote to Fermat in July (**III 381**).[63] But at least until September 1654 it had also been a year of astoundingly varied and intense mental vitality. With Fermat, Pascal had created the probability calculus. Single-handedly he had made important advances in the geometry of indivisibles, pursued his work on conic curves and sections, contributed to the binomial theorem, constructed the Triangle, extended the applications of the Mystic Hexagon, and founded hydrostatics. And, as we know from Leibniz (**II 227**), there was much more of that abounding insatiable inventiveness than we can perhaps even imagine: the drafts, jottings and intuitions of a man too ill to organize his scientific work methodically. Especially in geometry, what has come down to us on the printed page is probably only a tiny part of Pascal's total intellectual achievement.

Gilberte and Florin Périer were unable to be present at Étienne's deathbed, nor at his funeral. She was expecting her fifth child: a son, Louis, born on 27 September 1651. Pascal, writing to them (490–501) on 17 October 1651 about their father's death, was inspired to rare heights of eloquence and sensitivity: this letter is a most moving meditation on the significance of death. To him the distinction is fundamental between nature and grace, reason and faith. Those philosophers who have sought to found their lives and teaching upon the precepts of reason alone, and that includes almost all, have gone woefully astray: for 'there can be no doubt that Seneca and Socrates have said nothing persuasive on the subject. They have been in the same state of error as blinded all men in the beginning: they have all looked upon death as being natural to man; and all the discourses they have based on this fallacious principle are . . . futile' (492). Through the grace imparted by the sacrament of baptism a man dies to the world into the fullness of spiritual life: 'we did not lose my father at the moment of his death. We had, so to speak, lost him as soon as he entered into the Church through baptism. From that moment he belonged to God. His life was dedicated to God: in his actions he was concerned with the world only for God's sake' (495).

Death would be an obnoxious thing if it separated the holiness of soul and body; instead, it is the liberation of the human soul from bodily sin. Men are reluctant to die because, in the beginning of things, the purity and innocence of Adam caused him to dread the ending of a life which had been lived in perfect obedience to God's law. The fall of Adam has extinguished this perfect innate obedience, without extinguishing primitive Adam's fear of death. In its sinfulness, and in the discord between soul and body which exists within every man, the human race should now rejoice at the prospect of death and deliverance. 'It is not fitting . . . that we should be without grief' over Étienne's death, 'like angels which have none of the sentiments of nature; but equally, it is not fitting that we should be without consolation, like pagans who have none of the sentiments of grace: but it is fitting that we should be both distressed and comforted, like true Christians, and that the consolation of grace should prevail over the sentiments of nature' (499). Here, therefore, the paradox and the contradictions are already present in embryo which Pascal will use to disarming advantage in the *Thoughts*. Man's natural view of his natural human life is not *ipso facto* the right one. The insights into life's mystery which he acquires through the exercise of his human reason, through observation of his fellow men and through probing self-knowledge are so far from being complete as to be positively misleading: the little that unaided reason can impart to us must be richly supplemented by the revelations of faith. 'Let us no longer consider the human body as if it were some infected carcass, for deceptive human nature conceives of it in this way; but rather as the eternal and inviolable temple of the Holy Spirit, as faith teaches us' (495).

Death is ordained by God, Who ordained that His own Son should die in the world. The whole of human life is a sacrifice tending towards God; and both the centre of that sacrifice, and the clue to its meaning, lie in Christ. 'Let us therefore consider death in Jesus Christ, and not without Jesus Christ. Without Jesus Christ it is horrifying, detestable and abhorred of nature. In Jesus Christ it is quite different: it is amiable, holy, and the joy of the faithful. All is gentle in Jesus Christ, even death itself; and this is why He suffered and died in order to sanctify death and suffering; and why, as both God and man, He has been both all that is great and all that is abject, so as to sanctify all things within Himself' (493). As with Jesus, so with every human being created in God's image. Through their father's death 'the will of God is accomplished in him, and his will is absorbed into God' (495). At various points in the letter Pascal is at pains to establish that the doctrines he is putting forward for the Périers' comfort are not specifically his but have come from more authoritative sources: 'what I have learned from a great man in the time of our greatest affliction' (492); 'I have learned from a holy man in our afflictions' (500); 'This is what I have learned from two very great and very holy persons . . .' (496) We may think here of Singlin and perhaps also some other priest in the Port-Royal circle (it could not have been de Saci but might, though improbably, have been Arnauld); and as for the doctrines themselves, their indebtedness to an Augustinian tradition mediated by Jansenism is obvious. Particularly reminiscent of St Augustine is the accent on the fallibility of human reason and on the total inadequacy of unregenerate human nature to attain salvation. 'Let us therefore illuminate the error of nature with the light of faith' (497): in this echo of Augustinian thought, as also in his notion of paradox (both of them present in the letter to the Périers), lie the very foundations of Pascal's future defence of the Christian religion.

The content of the Letter may indeed largely have been inspired by Singlin's sermons and a reading of Saint-Cyran and St Augustine; but the presentation of its doctrines, the polished balance of the individual sentences, the haunting fusion of persuasive earnestness and an insistence so repetitious at times as to become almost obsessive, are unmistakably Pascal's own. Moreover, there is one aspect of the Letter in which Pascal's originality as an exponent of Christian doctrine is already apparent: its emphasis on the Christ figure, not simply as an object of religious contemplation (for this was a commonplace in treatises of devotion) but as the incarnate evidence of a transcendent system of thought and experience. Indeed, it is this aspect which gives the Letter on Pascal's father's death its peculiar force and cogency.

His worldliness, in any case, was only ever worldliness in a strictly limited sense. It began as a scientific worldliness, the passion for intellectual renown, then gradually moved closer to the worldliness of fashionable society. To begin with, in April 1652 he was again drawn into the salons, exhibiting his mechanical calculator and performing his barometric experiments, expounding his theory of vacuums in lucidly simple terms for the non-scientist.

A little later he wrote to Christina of Sweden – at this time the foremost royal protectress of learning and the arts, the patroness of Descartes who two years earlier had died in her kingdom and her service – and received a reply from the Queen's doctor Pierre Bourdelot: 'You have the clearest and most penetrating mind that I have ever seen. With your assiduity in work, you will surpass both the ancients and the moderns, and you will transmit to those who follow you a marvellous facility in learning . . . You are one of those geniuses whom the Queen is looking for . . . She will be delighted to receive your machine and your address' (**III 27–8**). It was probably in June that the now perfected calculating-machine was sent off, together with a letter to the Queen of Sweden (502–4) in which Pascal roughly foreshadows for the first time the theory of discontinuous orders which later became so fundamental to the *Thoughts*: Christina is of exalted worldly rank, but equally exalted in the still finer rank of mental attainment; the two are discontinuous, and Christina is the first monarch in the history of the world to be sovereign both in worldly power and in intellectual distinction; 'for I have', he adds (503), 'a quite particular reverence for those who are exalted to the supreme degree, whether it be in power or in knowledge. Unless I am mistaken, the latter no less than the former can be considered sovereigns'. But the scientific demonstrations and the despatch of the calculator were merely replays of already established achievements, flattering Pascal perhaps with the pomp of Courts and the dazzle of ducal salons.

It is not easy to picture Pascal's social manners in the early years of his life. He seems as a young man to have had something of the brusqueness and awkwardness of the provincial Auvergnat. His precocious scientific and mathematical bent – as can so readily happen in pure scientific research – had not been accompanied by any corresponding interest in the humanities, any finer flowering of civilized sensibility. To the end of his days, in fact, Pascal was never closely interested in music, nor did he become a connoisseur of painting. He likewise disdained the theatre and ignored poetry. Although a contemporary of Lully, Charpentier, Claude, Philippe de Champaigne and Poussin, he appears to have remained almost sublimely indifferent to their achievements. Indeed, he later censured painting for its vanity (40*). But though, certainly in his earlier years, he was not cultivated in the widest sense, this did not mean that he was indifferent to the relish of fashionable living. For over two years, and particularly from September 1653 to September

1654, he was on the fringe of the smart, carefree existence of the wealthy and socially distinguished, though – if only for financial reasons – never entirely belonging to it. Cultivating in their company a knowledge of the ways of the world, he gained a remarkably perceptive understanding of their outlook, mentality and attitudes. To the perhaps not entirely impartial Gilberte (10), he acquired 'the ways and airs [of the Court] with as much pleasure as if he had been brought up in it all his life'.

These years, to all students of Pascal, are known as his 'worldly period'; and it is a striking comment on the civility of the French, both in the seventeenth century and since that time, that the very words *mondain, honnête homme, bienséances, l'art de plaire, galanterie* and *talents d'agrément* are in their fullest connotations untranslatable. The world into which he had moved was little different in essence from that which, a hundred years later, Rousseau found so uncongenial and to which he singularly failed to adapt. Adaptability was indeed that society's most highly regarded virtue. The *bienséances* were the norms of conduct which were acceptable within that charmed circle, 'proprieties' which Corneille and Racine in their tragedies were at pains never to offend. The *talents d'agrément* were those gifts of social intercourse which, even if to some extent endowments of Nature, nevertheless had to be carefully nurtured. It was above all imperative for the *honnête homme* – the gentleman, the man of taste, dignity and good breeding – to adjust himself, both in his words and actions, to the outlook and susceptibilities of those with whom he came into contact. *Galanterie* was the refined expression of this ideal in the relations between the sexes, at times perhaps a little 'precious', exaggerated or insipid but fundamentally the ritualizing of a romantic courtesy. Whether sexually or in the more commonplace intercourse of society, the *art of pleasing* was central to the aspirations and actions of that élite.

Not only did Pascal regularly attend drawing-rooms, thereby acquiring something of the polish of a man of fashion, he also developed closer and deeper friendships with men to whom *honnêteté*[64] was an ever-present ideal of life. He now came into renewed contact with the Duc de Roannez, close to whom he and his family had lived in the Rue Brisemiche in the years between 1635 and 1639, and in the duke's circle met such perfect embodiments of civility as the Chevalier de Méré, a knight of Malta, and the wealthy bourgeois Damien Mitton: he thus became part of a tradition of courtly behaviour extending back to Castiglione's *Il Cortegiano* and beyond. Méré, indeed, was a latter-day Castiglione in that he too has recorded his philosophy of courtliness for posterity. Just as Castiglione, in early sixteenth-century Rome, felt that the height of civilization – in that most civilized of contemporary societies – lay in good manners, so Méré (in common with much of the thinking of the *Précieuses*) believed that a refinement of social attitudes would exert a tonic effect on the life of France, in both its thought and action. 'If anyone were to ask me what *honnêteté* amounts to', wrote Méré in his posthumously published *Concerning True Courtliness*, 'I would say that it is nothing other than to excel in everything that relates to the *agréments* and the *bienséances* of life . . . I do not understand how anything beneath Heaven can be above *honnêteté*; it is the quintessence of all virtues . . . The society of gentlemen is worthy of cultivation; but conversations with ladies, whose gracefulness puts us in mind of the *bienséances*, are more necessary still if we are to become perfectly accomplished in *honnêteté* . . . Perfect *honnêteté* is expressed when we choose the best ways of living happily, and of making those happy who are worthy of it'.[65] What he stressed above all was that the true *honnête homme* will always display such sensitivity to the feelings of others as to adapt with perfect empathy to their ways and outlook, whatever the circumstances.

Leaving aside a dubious investment, the consequences of Pascal's 'worldly period' were threefold. The insights which it afforded him into the urbanity and sophistication of polite society are obvious. Secondly, it brought him into close and repeated contact with *libertinage, incrédulité* – or what would nowadays be called agnosticism. And, more immediately important, the friendship with Méré (if friendship it was!) also had immense consequences for his mathematical work.

Although the term itself was coined by a scientist (by the zoologist T. H. Huxley, in 1869), agnosticism in the seventeenth century was rare in the extreme amongst scientists. At any rate, it was rare to find them openly professing it. Both Descartes and Newton believed in God: Descartes in God as a First Cause but not as the Supreme Good, a Being Who had devised so effective a cosmic mechanism that it left little if any room for Providential control; Newton, on the other hand, proclaimed his belief in a more voluntarist God.[66] Many scientists, mathematicians and philosophers – for example, Mersenne, Malebranche, René de Sluse, Antoine de La Loubère, Jean Picard and John Wallis – were actually priests of the Roman Catholic or Anglican churches, though at least one man who was a priest in name (Gassendi) was overtly freethinking. Even Étienne Noël had taken a genuine, though somewhat misguided, interest in the natural sciences! But, if agnosticism was rarely professed by scientists, it was also rare for it to be voiced by the middle classes generally. Certainly, doubts about the existence of God were held and expressed in some quarters with a fair degree of openness: not, however, in Auvergne or in the God-fearing Norman circles in which Étienne Pascal moved or in the *académies* where he spent his intellectual leisure. Such bold pronouncements were usually confined to the rich and powerful, those who were respected and even feared by the ecclesiastical authorities for their social and political influence. Not until he had made contact with the 'worldliness' of Méré, Mitton and Roannez did Pascal really come to terms with agnosticism.

Religion from now on was to occupy a deeper and deeper place in Pascal's life, to the point of largely overtaking his scientific interests and irradiating his existence. Not that it was to be a steady, constant trend away from science, away from the world, onwards and inwards towards the devotional life. On the contrary, from a narrow devotional point of view (if not also from a spiritual one) it was to be a chequered path, with many reverses and many misfortunes. The period of Pascal's 'worldliness' was now beginning, and would rise to an intense climax during the twelve months from September 1653 to September 1654. To his sister Gilberte, in her austere moral rectitude so similar to Puritanism that the extremes of the Christian religion seem to meet, this period was the most wasted in his whole life (10). But what should have been clear to anyone except those for whom self-abnegating religion is the sole precept of human life was that Pascal – throughout his worldly vanities, and with all his attachment to ambition and renown – was becoming not only a profoundly devout churchgoer, a fervent reader of the Scriptures and of the spiritual exercises of Saint-Cyran and others, but a man with his own interpretive understanding of the relationship of religion to the world.

Echoing St Paul, and in terms verging upon Platonism, he could write: 'Corporeal things are but an image of spiritual things, and God has shown forth invisible things through things that are visible' (484);[67] and this was in a letter to Gilberte herself, dated 1 April 1648 (483–6), though something of its inner mysticism may have escaped her. Again, in the same letter: 'We must think of ourselves as criminals in a prison overflowing with images of their Redeemer'

(484). Sinfulness, he asserted, is 'mistaking the image for the reality' (485). Already taking shape in his mind were thoughts which foreshadow the future development of his arguments in defence of Christianity.

Pascal was approaching the supreme crisis in his spiritual and intellectual life. The mental ferment which had found various expressions in the work on probability, the arithmetical triangle, binomial coefficients, indivisibles (the infinitesimal calculus in embryo), hydrostatics, the theory of number, magic squares and conic sections had also, it seems, been leading mysteriously towards a dynamic mystical experience. Mysticism – an intensely personal subjective experience of the Godhead, direct intuition of the pure truth of being, a localized sensation of the Divinity which (in St Paul's words)[68] 'filleth all in all' – is an experience given to few. Pascal had known nothing of it in his 'first conversion' in 1646; nor had his father. Nor have the vast majority of those who throughout the ages have called themselves Christians, or Hindus, Moslems or Jews, and who have striven against all temptation to observe either the Sermon on the Mount or the precepts of the *Vedanta*. 'What we sow in the minutes and spare portions of a few years', wrote Pascal's contemporary Jeremy Taylor, 'grows up to crowns and sceptres in a happy and a glorious eternity'; and this approach has been the spiritual touchstone of most believers.

To a few men and women, however, the experience of the Divine has been infinitely more profound, intimate and visionary. In the words of St John of the Cross, mysticism is 'a heightened sensation of the divine essence'. 'My eye and God's eye', wrote Meister Eckhart, 'are one eye, one vision, one recognition, one love . . . God has begotten me from eternity that I may be Father and beget Him who begat me'.[69] St Theresa of Avila, in one of her visions, had the feeling that she was a mirror without a frame and without dimensions, and with Christ shining in the centre of it; and that the mirror itself – she did not know how – was in Christ. To Henry Vaughan Eternity was

> Like a great ring of pure and endless light,
> All calm, as it was bright;
> And round beneath it, Time in hours, days, years,
> Driv'n by the spheres
> Like a vast shadow mov'd; in which the world
> And all her train were hurl'd.

'If every earthly pleasure were melted into a single experience and bestowed upon one man', said Ruysbroeck,[70] 'it would be as nothing when measured by the joy of which I write . . . The body itself can know no greater pleasure upon earth than to participate in it; and there are moments when the soul feels that it must shiver to fragments in the poignancy of this experience'. 'Were one drop of that which I am feeling', said St Catharine of Genoa, 'to fall into Hell, all Hell would become life eternal'. Even to the greatest mystics, however, the mystical experience is a rare one: it is dazzling yet transitory, and perhaps never returning. During the time it lasts, which may be only a few minutes (St Theresa did not believe it was 'ever so long as half an hour'), there may be the sensation that time and space are abolished, and a transcendence of the distinction between good and evil such that, in the words of Dame Julian of Norwich, 'all manner of things shall be well'. Then the imperfections of the world subside, the problem of the existence of evil within a God-created universe loses its nagging insistence, and the transcendence of the Eternal Almighty is suspended as the 'immortal, invisible' God takes on the form of a distinctly perceived image within a particular human soul.

Pascal likewise was now about to undergo two hours of such intense spiritual ecstasy, such close communion with the Divine, that the whole of his life up to this moment would seem only the pale foretaste of all the tumultuous joys, yearnings and sufferings which suddenly became his. This was his one true conversion, the complete turning-around of his life to the dedicated service of God. Throughout his thirty-one years he had believed in his religion, and all the more passionately during the last eight. Now the almost unspeakable bliss of his *mystical* conversion would cause him, whilst still remaining a pre-eminent mathematician, to focus his finest energies – scientific, spiritual, forensic, literary – in the service and the championing of a God in Whom he no longer merely believed but Whom he knew and had seen. His spiritual motto was to be taken from the words of St Paul to Timothy: *Scio [. . .] cui credidi*: 'I know Whom I have believed'.[71]

3

Jansenism

Port-Royal des Champs, the seat of Jansenism, stood (until razed to the ground in 1711) about eight miles south-west of Versailles, in the Chevreuse valley, close to the village of that name. It is a spot which even today is renowned for its picturesque beauty. Wooded slopes on all sides run down towards undulating meadow land through which winds a small river. Nowadays much of the area is sparsely populated with either the evening or weekend retreats of Parisian businessmen; in 1655, in its beauty, tranquillity and stillness, it must have been peculiarly fitted to spiritual retreat and meditation. But Port-Royal was also low-lying and damp, and considered by many to be physically unhealthy.

A Cistercian nunnery had occupied the site since 1204. But the abuses and spiritual laxity of monasticism in the late Middle Ages, which were a major factor contributing both to the dissolution of the English monasteries and to the Reformation generally, also took their toll at Chevreuse. Not even the unparalleled austerity of the Cistercian rule could withstand the decline. The Wars of Religion hastened and completed this process, with the result that by the end of the sixteenth century the convent, both spiritually and materially, was at its lowest ebb. Into this lamentable situation there then came two sisters, and with them a whole family, who not only restored a flourishing religious community in the valley, more devout than anything that had been seen under the Cistercians, but who also created a spiritual movement, both of the heart and mind, which radiated across France and (through Pascal) throughout the world.

The elder of the two sisters, Mère Angélique Arnauld, was no more than eleven years old at the time of her election as Abbess of Port-Royal. She was the daughter of a distinguished but somewhat impoverished lawyer, Antoine Arnauld, who in 1594 had achieved notoriety by a speech accusing the Jesuits of complicity in the assassination of Henri III, murdered five years previously by a Dominican friar: this was to be only the first of the Arnaulds' brushes with the Jesuits. Having a family of four sons and six daughters (not to mention the ten other children who did not survive infancy), Antoine – like many a father in Molière's comedies – was well content to place his daughters in convents, so avoiding the irksome necessity of finding sufficient dowries. The second and third daughters, Jacqueline and Jeanne, were both provided with suitable niches at ages exceptional even by the lax standards of the early seventeenth century. At seven years of age, in 1597, Jacqueline became Abbess Coadjutrix (i.e., with the right of succession) at Port-Royal, thus assuming in tenderest infancy the title of *Mère* Angélique. Still more remarkably, Jeanne became Mère Agnès, Abbess Coadjutrix of Saint-Cyr (another Cistercian convent), at the age of five and a half.

It was on the death of the then Abbess that Mère Angélique succeeded her at Port-Royal in 1602. To her parents this appointment was little more than a sinecure, an easy way out of the impossible problem of endowing her suitably for marriage. And so it also seemed to Angélique, at any rate during the first six years of her superiorship. In 1608, however, at the age of eighteen, she too had experienced a conversion. After hearing a Capuchin friar preach movingly in her Chapel on the humility and poverty of Jesus, she became a reformed woman; and so set to work to reform her community. Her parents meanwhile still looked on Port-Royal as something of a secular enclave, with an Abbess still subject to their parental authority. But their illusions were shattered in September of the following year, when she bolted and barred the doors of her convent to keep out her parents and family, being (as she claimed) subject only to the General of the Cistercian Order. There could be no clearer signal that she had now repudiated the fashionable view of conventual life: unmistakable but brutal, such a signal was perhaps unavoidable in that she was confronting parental authority. The earnestness of her new conception of her duties was strongly reinforced by St Francis of Sales, her spiritual director from 1619 until his death in 1622: St Francis, through his preaching, his missionary zeal, and the urgent simplicity of his *Introduction to the Devout Life*, made as enormous and persuasive an impression upon her as he had also made upon Jeanne de Chantal with whom he founded the Order of the Visitation. It was he who gently chastised Mère Angélique with the remark (much beloved of Ronald Knox) that in any religious mission 'it might be better not to catch just a few big fish, but a larger number of smaller ones'.[1]

The establishment of an alternative religious house in Paris, in the Faubourg Saint-Jacques, dates from 1626 and was decided upon, by

Mère Angélique, on account of the damp swampy climate of Port-Royal des Champs. The convent in the valley of Chevreuse was closed, and the whole community moved to a fringe of countryside on the outskirts of Paris, just south of the spot where a Benedictine monastery, the Val-de-Grâce, was soon to be built by Anne of Austria. Sébastien Zamet, the pious but eccentric Bishop of Langres, became the convent's new spiritual director. Within a year of Port-Royal's removal to Paris Zamet had persuaded Pope Urban VIII to transfer the community from the Cistercian Order into the direct care and control of the Archbishop of Paris. Not content with having secured this unique relationship, Mère Angélique (who had already sought, vainly, to be admitted to the Order of the Visitation and who for a spell had been abbess of a convent near Pontoise whilst Mère Agnès replaced her in Paris) now enlisted Zamet's help in establishing a new religious order, the Institute of the Blessed Sacrament, for which she left Port-Royal in 1633. But the new religious foundation did not prosper, and she returned three years later as Novice Mistress to the community of which Mère Agnès had again become Superior. In 1638 Mère Angélique took over from her sister as Abbess of Port-Royal for a third time. Like the by now defunct community which she and Zamet had established together, Port-Royal was dedicated to the perpetual adoration of the Blessed Sacrament; and the nuns' habit became the white scapulary with a red cross vividly portrayed to posterity in the 'Ex-Voto', the 'Portrait of Mère Angélique' and other works of Philippe de Champaigne. In 1648, by which time Pascal and his sister had returned from Rouen to the Rue Brisemiche and Jacqueline was paying regular visits to the Faubourg Saint-Jacques, the increasing number of postulants, novices and nuns compelled Mère Angélique to reopen the conventual house at Chevreuse; and there was constant movement – of superior nuns, postulants, priests and well-wishers – across the twenty-three miles that separated Port-Royal des Champs from Port-Royal de Paris.

But, though Mère Angélique was Abbess of Port-Royal three times and Mère Agnès twice, they by no means represented the full complement of Arnaulds within that community. The mother of these two ladies became a nun at Port-Royal on becoming a widow. So, too, did their married sister, Mme Catherine Lemaître, known in religion as Sister Catherine of Genoa. Three further sisters, Anne-Eugène, Marie-Claire and Madeleine de Sainte-Christine, were also nuns of Port-Royal. Likewise in the community were the six unmarried daughters of Mme Lemaître. Nor was this the end of it. For the peculiar originality of Port-Royal lies in the fact that it was very far from being merely a nunnery. Port-Royal, apart from occupying a split site from 1648 onwards, was in fact three distinct though interconnected entities: a convent, a residence of *Solitaries*, and a school. And because of the fact that the country part of the establishment was also a residence of Solitaries and a school, Port-Royal also found room to accommodate six male members of the Arnauld family.

At the height of its fame and prosperity there were probably about thirty Solitaries at Port-Royal des Champs, though the population was fluctuating and numerically unstable and could never be predicted with certainty. Amongst these thirty were two brothers of Mère Angélique, the most controversial of living theologians Antoine Arnauld and his more diplomatically inclined courtier brother Robert Arnauld d'Andilly; there were also the three sons of Mme Lemaître (nephews of Mère Angélique): Antoine Le Maître, Simon Le Maître de Séricourt and Isaac-Louis Le Maître de Saci. Yet another Arnauld Solitary was Luzanci Arnauld d'Andilly, Robert's son, who in fact joined the community before his father. Isaac-Louis de Saci was a priest and theologian, Robert Arnauld d'Andilly a gardener and a lover of fruit-trees, Simon de Séricourt an ex-soldier, and Antoine Le Maître a gifted young lawyer whose renunciation of an exceptionally promising career may well have inspired the theme of Corneille's tragedy *Polyeucte*. Though not of the religious community of the Blessed Sacrament, they lived close to it. The convent itself lay in the valley, but their home was Les Granges, originally the farmhouse from which the agricultural land of the community was managed. As the number of Solitaries grew, extensions were built to the house whilst one or two of these holy men occupied independent *cellules* – huts or tiny cottages – about the estate.

The running of the farmland was just one of their concerns. Each of the Solitaries (when he was in residence: for they could come and go as they pleased) had his own self-allotted tasks. Whilst some gardened close to Les Granges, others helped to drain the stagnant swamp at the foot of the hill. It was a monastic community without formal vows. Like any such community, it was composed partly of priests and partly of devout laymen. Like monks, they chose a life of withdrawal, penitence and prayer, though without feeling either the vocation or the necessity of a particular religious order. Their spiritual development, intellectual fellowship and collective labour for the material well-being of Port-Royal des Champs were untroubled by personal financial anxieties. Such money as they had, they had made over, though not given, to the community in return for a pension.

Under Antoine Arnauld's influence, their free time in each other's company was spent in intense metaphysical discussion. Arnauld had become fascinated by Descartes's theories of the sharp dualism of soul and body, mind and matter, as expounded in the *Discourse on Method* and the *Meditations on Primary Philosophy*. Nicolas Fontaine has left a graphic description of these debates in the preamble to his account of Pascal's *Conversation with M. de Saci*. Descartes had maintained that animals are terrestrial machines, automata devoid of either mind or soul, and totally distinct from human beings in that the latter, though animal-like in their bodily organization, are additionally equipped with a God-given 'rational soul'. At Port-Royal des Champs, Fontaine recalls, there was hardly a Solitary who was not discussing automata. 'No one worried any more about beating a dog; they were quite unconcerned about giving it a good hiding, and would scoff at people who felt sorry for these animals as if they had been capable of feeling pain' (**IV 28**). Echoing the very term used by Descartes in his *Discourse on Method*,[2] the Solitaries would say that dogs 'were only clocks; and that the yelps they made on being beaten were only the sound of a little spring that had been set in motion, but that the brutes were quite incapable of feeling. They would nail such poor animals on to a plank by their four paws, then open their insides whilst they were still alive, to study the circulation of the blood which was a keen talking-point'. (Descartes had pronounced in favour of William Harvey's theory of the circulation of the blood through the arteries and veins, whilst denying that it was driven round the body by the contractions of the heart.)[3]

When not engaged in a form of scientific enquiry which from a post-Freudian viewpoint comes perilously close to sadism, the Solitaries' intellectual energies were abundantly employed in medicine and teaching. One was a surgeon (562), another – Jean Hamon – a doctor who cared not only for his fellow Solitaries but also for the convent and school. The school which Saint-Cyran had founded in 1637 in the Faubourg Saint-Jacques was now also quartered in an extension of Les Granges and elsewhere at Le Chesnay and Les Trous,[4] both near Versailles; these adjuncts to the school were known as the Petites Écoles and by 1655 numbered about fifty pupils. Not the least aspect of the Solitaries' work was the fresh and stimulating impetus which they and their followers gave to schoolteaching. Nicole and Claude Lancelot and others taught ceaselessly at Les Granges. Books written by the Solitaries or their friends, Latin and Greek primers, treatises on logic and grammar, were produced on their own printing-press. In 1660 Arnauld himself drafted a geometry textbook,[5] to which Pascal contributed the ideas expounded in *On the Geometrical Mind* (576–92). The latter's nephew, twelve-year-old Étienne Périer, was a pupil at Le Chesnay (1378) at the time of his uncle's first visit to Port-Royal des Champs. Another pupil at the Petites Écoles, from April 1658 until the summer of 1660, was Charles II's eldest illegitimate son the Duke of Monmouth.[6] But by far the ablest alumnus was Étienne Périer's contemporary Jean Racine, the future playwright in whose tragedies *Phædra* and *Athaliah* Jansenist thought patterns are often detected.

For all the brilliance of the lay Solitaries, and the forthrightness and strength of mind and heart of Mère Angélique, it was the priests of Port-Royal – Arnauld, Singlin, Saci, Nicole, Rebours – who exerted the predominant influence. Before the small and unassuming Singlin even Mère Angélique would quail. Arnauld's cause was espoused by the whole of Port-Royal, and it was this cause which led to their downfall. Pierre Nicole was the schoolmaster *par excellence*, Singlin the confessor, Saci the man of moderation and the scholar who went back again and again to St Augustine's *City of God, Forgiveness of Sins and Baptism, Nature and Grace* and *Human and Divine Righteousness* not looking for 'new arguments with which to argue well, but for what might give new sustenance to his piety' (**IV 27**). But there was another priest – Saint-Cyran – whose influence twelve years after his death was still stronger in the mind and memory of Mère Angélique than any other's. She had first come into contact with him in 1621[7] (though her eldest brother Robert Arnauld d'Andilly had met him even before that)[8] and he had been of great help to her in the later 1620s in her attempts to set up the Institute of the Blessed Sacrament. On leaving Port-Royal to become Superior of this institute in 1633, she appointed Saint-Cyran to be its spiritual director; returning with her to Port-Royal in 1636, he replaced Sébastien Zamet as its director, founded the school, established the Solitaries on the model of St Antony of Egypt, St Pachomius and other hermits, and imparted a new rigorism (some have said: self-importance) to the community both through his preaching and writing. The influence he exerted was of a mysterious, almost magnetic kind, which Richelieu's four-year imprisonment of him without trial only served to glamorize with a suggestion of martyrdom. Richelieu had noted his powerful evangelizing effect, and justified his treatment of Saint-Cyran on the grounds that if the same had happened in good time to both Luther and Calvin Western Europe would have been spared much turmoil. Saint-Cyran, in other words, was a reformer – both of hearts and of the Church. It is an accident of history that, from his youth onwards, he was a close friend of Cornelius Jansen. Even in their student days at Louvain they had dreamed dreams and schemed schemes together for the reform of Catholicism. But the reform which both envisaged was a *reactionary* reform taking the Church back to the pristine ideals of St Augustine and Patristic theology.

Ernest Mortimer distinguishes between 'Jansenism', 'Saint-Cyranism' and 'Arnauldism'. 'Saint-Cyranism alone', he writes,[9] 'without the Jansenist polemic, might have retained an honoured if rather embarrassed welcome within the Church'. In his view, it is the admixture of Jansenist theology which makes Saint-Cyran's moralizing and reforming rigour so unpalatable. Ronald Knox, though severe both on Saint-Cyran and the Arnaulds, is much harsher concerning the latter. 'To be a Jansenist', he urges[10] (echoing Bremond), 'you must always be writing against somebody. A theologian by trade, Arnauld was a barrister by instinct. . . '; Arnauld's 'example poisoned the very roots of charity at Port-Royal';[11] Mère Angélique he sees as 'incurably self-conscious', 'always dramatizing situations';[12] the entire community of nuns and Solitaries was, he asserts, 'Little Gidding reduced *ad absurdum*'.[13] But this last epithet is surely unjustified. Just as the family of Nicholas Ferrar worshipped at Little Gidding in their community, so the Arnaulds did likewise – though without the sense of stability and established authority of Laudian England, and without the blandness and suavity of the High Anglican Church. The fact that the Arnaulds were pugnacious whereas the Ferrars were not was entirely due to the circumstances of their history. There is a holiness, rectitude and dignity about Mère Angélique and Singlin and Nicole which is different from the Ferrars' holiness but holiness nevertheless.

The Arnaulds were the expression, and indeed the creators, of a twice reformed[14] religious community at Port-Royal just as the

Ferrars were the embodiment of theirs – which, arising spontaneously, had never been in need of reform. As a concept, 'Arnauldism' does not exist. Without its Jansenist content 'Saint-Cyranism' would have no substance. Even as it is, 'Saint-Cyranism' represents nothing more than the bravest attempt – first of one teacher, then of one receptive religious community – to translate dogmatic precept into practical moral action. Port-Royal des Champs is only more *absurd* than Little Gidding in that its Jansenism inherited through Mère Angélique from Saint-Cyran was less gentle, more stubborn and more reactionary than the religion of the Ferrars.

Cornelius Jansen, the friend of Saint-Cyran from their undergraduate days at Louvain, was rector of the episcopal college at Bayonne when Saint-Cyran was a canon of the cathedral in that city. During the five years in which they shared lodgings in Bayonne, every free moment of their time was devoted to a study of the early Fathers of the Church: Clement of Alexandria, Justin Martyr, St Ambrose, St Jerome, Pope Gregory the Great, but most notably St Augustine. In 1617 Jansen returned to the University of Louvain where soon afterwards he became a doctor in the Faculty of Theology. There he remained in growing favour for nineteen years, having in the meantime been appointed to a chair in Biblical exegesis. Whilst Saint-Cyran was the pastor and director of souls, Jansen was the writer and thinker. In the last two years of his life, however, until his death in 1638, Jansen too occupied a pastoral and administrative position as Bishop of Ypres. His untimely end – perhaps the result of his trying to combine searching scholarship with the arduous duties of managing a diocese – prevented him personally from publishing the extensive study of Augustinian thought on which he had been labouring for twenty-two years.

The *Augustinus*, the source of so much embittered controversy in the next twenty years, was posthumously published in 1640. Even before its publication by the University of Louvain, the Jesuits were making strenuous efforts to prohibit it. Designed as an exposition of St Augustine's doctrines of grace and regeneration, and to refute the heresies of Pelagius, its orthodoxy (on the surface) could scarcely have seemed more respectable. St Augustine, after all, was not only a Father of the Western Church but (with St Jerome, St Ambrose and St Gregory the Great) one of its four 'doctors'; his *City of God*, perhaps the supreme exposition of Christian doctrine. Pelagius's teaching that man did not need God's grace in order to achieve his salvation, that original sin did not exist, and that all that a man needed to do to win salvation was to exercise his freewill wisely: all this had been roundly denounced, and declared heretical, at the Synod of Carthage in 416, and St Augustine in no less than fifteen treatises had taken a foremost part in the denouncing. Likewise, Augustine had been involved in the denuncation of two other doctrines, Manicheism and Donatism, both of them declared heretical. For well over a thousand years Augustine had been the bulwark of authority and dogmatic wisdom within the Church, the pure if rigorous teacher systematizing and coordinating Christian doctrine and linking Gospel times to the present.

The idea that a man could only be saved by the arbitrarily bestowed grace of God, independently of his own human freewill, was however a somewhat rebarbative doctrine essentially inspired by Augustine's own particular and individual religious experience. Not until he was a grown man of thirty-one years of age, and the father of an illegitimate son, was Augustine suddenly converted. This impulse of faith and self-surrender always appeared to him subsequently as God's will rather than his. Why, if it had been in his power to choose, did the conversion come at one moment rather than another? and, because therefore God had willed it so, His act was an arbitrary bestowal of grace since the same man whom He had refused to save in adolescence He had saved in his maturity.

Pelagius, on the other hand, had considered it both illogical and unjust that infants dying unbaptized should be condemned to perdition because of some notional Original Sin, that holy men happening to die before the birth of Christ should also be condemned to damnation, that on God should be heaped the blame for the shortcomings of man, and that throughout the Church men should be encouraged to rely for salvation on the reciting of creeds and near-magical efficacy of sacraments rather than on cultivating a Christian character. Augustine, however, had his millennium of triumph and Pelagianism vanished almost without trace until the sixteenth century.

The erosion of the old medieval complacencies, the Reformation, the Counter-Reformation, the emergence of the Jesuits changed this theological situation entirely. As early as 1567, and again in 1579, Baius's comparatively mild expression of Augustinianism had been condemned by Popes Pius V[15] and Gregory XIII.[16] In 1588 the Spanish Jesuit theologian Luis Molina took a further stride in this direction with the publication of his *Reconciliation of Freewill with the*

Gifts of Grace, a book which inclined strongly towards the Pelagian doctrine of man's natural remediability and the possibility of salvation through faith and good works. To the Jesuits, seeking to regain many countries of Western Europe for the Catholic faith, it was imperative that the harshness and severity of Augustinian teaching – the seeming hopelessness of the human predicament where God had not chosen to intervene – should be watered down. In order to win converts, they were only too ready to stress that God's yoke is easy and His burden light.

Reacting against these developments, Jansen wished to restore the Augustinian tradition to its former position of supreme orthodoxy. This posed a difficult problem for the Church and especially for the Jesuits, alarmed even at the first rumour of the book's appearance. No sooner had it been published (its popularity immeasurably increased by all the efforts to ban it) than the Jesuits pounced to destroy. What they found in the three folio volumes both appalled and delighted them. They were appalled by the unmitigated severity of Jansen's teaching, but they were not without a certain satisfaction that such unrelieved bleakness would make it a fairly easy target.

The doctrine expounded by Jansen as the true and traditional teaching of St Augustine included the following tenets: that the natural human will, in its perversity, was incapable of doing good; that to do good, and to be saved, a man needed God's grace (natural man, deprived of grace, being unable even to turn to God); that the soul was condemned to eternal damnation if it was not sustained by grace; that grace was not given to all men, but only to the elect; that, man not being free to win his own salvation, it was decided by God whether or not he would be saved; that Jesus did not come into the world to save all men, but only for the small number of the elect; that only the elect were capable of fulfilling God's commandments but that to do this, even to pray, they needed an *efficacious grace* which would determine their will to obey God, or to pray; and that this efficacious grace, being wholly dependent on God's mercy, was not always given even to the elect but that, whenever it was given, it was irresistible: divine grace, therefore, was tantamount to Predestination. Only those on whom God bestows His grace can be saved, argued Jansen; and their salvation was foreordained.

It is doubtful whether even St Augustine himself believed in so black a picture of human nature as that painted – and somewhat arbitrarily selected from his writings – by Cornelius Jansen.

Augustine was an advocate, and advocates are easily led to exaggeration. He was also the foremost opponent of Pelagius, and it was precisely because of the threat of Pelagianism that he laid so much stress on Original Sin and on man's natural wickedness. It did not seem too difficult to the Jesuits to present Jansen's *Augustinus* as a warped, lopsided interpretation.

Pius V and Gregory XIII had already, in the late sixteenth century, forbidden anyone to write on the sensitive subject of divine grace without first obtaining their express permission. The publishers of the *Augustinus* had defied these constitutions. In the year following its publication, the book was banned by the Inquisition. The Jesuits were ordered not to raise the temperature of the debate by publicly canvassing the matter. In 1642 Pope Urban VIII forbade the reading of the *Augustinus* not only because it had been published without his permission but because it allegedly contained certain 'errors'.[17] Both in the Netherlands and in France the Papal Bull met with fierce resistance. The Sorbonne disputed its accuracy and authority. Nevertheless, Isaac Habert – a canon theologian of Notre-Dame, later to be appointed Bishop of Vabres – delivered a course of three sermons, on the first and last Sundays of Advent 1642 and on Septuagesima 1643, sternly denouncing the *Augustinus*. Antoine Arnauld, already well known for his book *On Frequent Communion*, countered Habert's arguments in his *Apologia [or Defence] of M. Jansen*, published in 1644; Habert replied almost immediately in his *Defence of the Faith of the Church*; not to be outdone, Arnauld prolonged this war of words in a *Second Apologia for M. Jansen*, published a year after the first.

The gist of Arnauld's argument, in both tracts, was that nowhere in *In Eminenti* had Urban VIII pointed to Jansen's precise violations of Catholic orthodoxy. No passages of the *Augustinus* were cited, no specific doctrines referred to. Short of accepting the automatic infallibility of Papal pronouncements (and that did not come until 1870), here was a serious flaw in the case presented by the Holy See. Arnauld, on behalf of the friend of Saint-Cyran with whom all his family's sympathies lay, had won a notable victory.

The dispute then centred around the extraction of propositions. The Jesuits clearly saw that it was essential to designate the precise heretical content of Jansen's work. First, in 1646, Habert produced a list of eight unorthodox statements allegedly contained in the *Augustinus*; these were to be sent to Rome. After much discussion, extending over three years, Habert's collection of eight supposed heresies was supplanted by the Seven, subsequently *Five Propositions* of Nicolas Cornet (syndic of the Theological Faculty of the Sorbonne), and these Five Propositions were sent off in the name of eighty-five French prelates for judgment and condemnation.

Cornet's eventual catalogue of Jansenist enormities ran as follows:

> 1) some of God's precepts are impossible to the just, who wish and strive to keep them, according to the present powers which they have; the grace, by which they are made possible, is also wanting;
> 2) in the state of fallen nature one never resists interior grace;
> 3) in order to merit or demerit in the state of fallen nature, freedom from necessity is not required in man, but freedom from external compulsion is sufficient;
> 4) the Semipelagians admitted the necessity of a prevenient interior grace for each act, even for the beginning of faith; and in this they were heretics, because they wished this grace to be such that the human will could either resist or obey;
> 5) it is Semipelagian to say that Christ died or shed His blood for all men without exception.[18]

Tortuously involved as they may seem to the late twentieth-century reader, Cornet's Five Propositions presented Urban VIII's successor Innocent X with a distressing dilemma. There was an undoubted ring of Calvinism about them; yet if they were to be found in St Augustine, they were also presumably 'orthodox'! What he almost certainly could not have known was that all five Propositions, whether or not they are present in St Augustine's writings, are not enshrined in the *Augustinus* anyway: they were distillations of, though not actually to be found in, Jansen's book.[19] The Pope would much have preferred the acrimonious argument to subside quietly of its own accord. Finally, however, he condemned[20] the first four of Cornet's Propositions as heretical, and the Fifth as false, rash and scandalous – adding that, if the exact meaning of the Fifth Proposition was that Christ died only for the elect, then it was also impious, blasphemous, contumelious, dishonouring to divine piety, and heretical.

But it was no part of Innocent's intention to impugn Augustinian doctrine; and *Cum Occasione* made this clear. Matters had reached an uneasy stalemate. The Jansenists – which meant the Port-Royal community – realized that there was only one line of defence left to them. It was one thing to condemn the Five Propositions as heretical (and to this verdict, emanating from the Pope, they reluctantly assented); but were these Propositions Jansen's? A summary could betray the meaning of the whole. Even a literal quotation, taken out of context, could be positively misleading. What was the meaning and message of Jansen: a subjective impression variously interpreted? How, then, could it be condemned?

The debate was straying into the higher realms of semantics, far away from the practical necessities, as Rome and the Jesuits saw them, of presenting Catholic dogma to the agnostic or Protestant in a favourable light. Innocent returned, therefore, to the attack in September 1654, declaring that the Five Propositions were actually to be found in the *Augustinus*,[21] and that they were an adequate summary of Jansen's meaning and should be condemned. Jansen had always foreseen that parts of his book might not be acceptable to the Holy See, and in this event would have preferred them to be expurgated from the MS of the *Augustinus* before publication (for this, he had given express permission in his will). Now, however, within fifteen years of the book's appearance, it had been condemned outright by the Pope more for the bleak hopelessness of its 'Five Propositions' than for their technical unorthodoxy. These, whatever the semantic nuances of the debate, certainly corresponded to the spirit if not the letter of Jansen's teaching.

Such was the situation of Jansenism at the time of Pascal's first arrival at Port-Royal des Champs: an orthodoxy which, for reasons of expediency, was ceasing to be quite so respectable and orthodox. In the interests of the Counter-Reformation, Augustinianism was being buried by the Jesuits. The Rouvillism of Pascal's 'first conversion' was not Jansenism; and, though his sister Jacqueline was a Jansenist nun and his nephew Étienne a pupil at the Petites Écoles, he was not actually a Jansenist himself. Admittedly, he appeared to M. de Saci on this first visit to be 'in accordance with St Augustine in all respects' (562); at the same time, he had not yet read the Fathers of the Church, though – by the light of his own reason – he is said to have come to a position analogous to theirs (561); he was, in fact, practically unversed in theology. Antoine Arnauld, on the other hand (with whom Pascal now came into close contact), was the most committed and belligerent defender of Jansen. For him there was only one way ahead. This was to go on emphasizing the distinction between what purported to be by Jansen and what was actually by him, between *droit* and *fait* (*de jure* and *de facto* condemnations): a lawyer's distinction rather than a theologian's, but then Arnauld had both the legalistic mind and the training of his father and grandfather. 'A theologian', as Ronald Knox has remarked,[22] 'should have been stiffer, a politician more pliant'. The pliancy would come in future from Pascal, the hard theology from the whole of the Port-Royal community; Arnauld would stick to his neat distinctions between *right* and *fact*.

4

A 'God-Inebriated Man'[1]

Pascal's genius and international scientific renown were well known to Isaac de Saci (560), who appears – from the account left by the latter's secretary Nicolas Fontaine[2] – to have been somewhat reluctant to take him on. Nevertheless, says Fontaine (561), he 'could not neglect seeing him, for decency's sake, especially when asked to do so by M. Singlin'. It is clear that Saci, like no doubt other members of the community, feared and disliked his 'worldliness'. 'The sacred light which [Saci] found in the Scriptures and the Fathers of the Church led him to hope that he would not be dazzled by all the brilliance of M. Pascal, who nevertheless charmed and entranced everybody'. What struck Saci most forcibly at the very outset of their discussions was that Pascal, though he had not read the Fathers of the Church (Augustine, Tertullian, Jerome, Irenæus, St John Chrysostom), was at one with them on all important matters of doctrine: 'with his penetrating mind he had found out all by himself the same truths which they too had found'. As yet unaware of the formidable background of centuries of theological history, he had a tendency – which others, less patient than Saci, might have found rather exasperating – to believe in the startling novelty of his religious views, convinced (perhaps from his scientific training) that he had somehow hit upon ideas so surprisingly original that no one before him had ever thought of them. In Saci's judgment, all the 'great things' said to him by Pascal were already to be found in St Augustine.

The *Conversation with M. de Saci,* recalling a discussion held in 1655 but not published until 1728, is important both as a statement of Pascal's general philosophical outlook and as foreshadowing his later development.[3] Embedded in the dense fabric of the discussion are two authors, one a philosopher and the other an essayist who (like Bacon) had deeply pondered the meaning of life, Epictetus and Montaigne.[4] Pascal began by informing Saci that these were the two writers whose views on the human condition he most deeply admired: Epictetus, he thought, was the best exponent of 'human duties' (562), the greatest philosopher of stoicism, whilst Montaigne was the ablest confuter of vain, self-stultifying human reason, the voice of pyrrhonism or philosophic doubt. Epictetus's greatness lay in the fact that he believed both in the sovereign power of God and in the human virtues of humility, resignation, self-denial and contentment. Montaigne, on the other hand, was unsurpassed in his demonstration of the relativity of things – the presumptuousness of men idly believing that their reason was capable of attaining objective truth. For the rest of his life Montaigne's *Essays* were perhaps to be Pascal's most haunting intellectual companion: not (in Ronald Knox's words)[5] 'to give him any depressing information about human existence', which 'he could read . . . himself, all too clearly, in his own sickly body, and his restless mind', but to reiterate again and again that all is uncertain, that atheism is foolishly arrogant to deny the existence of an Infinite but that likewise there is always as much to be said for an opinion as against it. Montaigne, Pascal remarked (564), 'imperceptibly destroys everything that is considered to be most certain amongst men, not to establish the contrary with a certainty which is his only enemy but merely to show that, appearances being equal on either side, you cannot tell what to believe'.

On Pascal himself such deep-rooted scepticism has a tonic effect, though in the generality of men he fears it would tend to induce lethargy (569, 571). 'From the basic premise', Fontaine writes (569–70), 'that outside faith all is uncertainty, and bearing in mind how many people there are who seek Truth and Goodness without making any progress towards peace of mind, [Montaigne] concludes that you must leave anxiety about these things to others; meanwhile remaining happy and unconcerned, gliding gently over things for fear of falling in if you press hard; and judging truth and goodness by first impressions, without probing too far, because they are so insubstantial that as soon as you begin to press them tight they escape through your fingers, leaving an empty hand'. But Epictetus also has his failings: laying before man an exalted ideal of nobility and duty, he does not recognize that the flesh is weak. He acts and teaches as if it lay entirely within man's power to do noble things, to perform acts of mercy, self-sacrifice, tolerance, industry, heroism – overlooking the pathetic weakness of the human will. The God in whom Epictetus believes, and whose existence he never doubts, is one who has implanted in human nature all that a man needs in order to lead a virtuous, dignified life (563). Indeed, in this view man is so God-like as to be almost on an equal footing with the Almighty; Epictetus even suggests that the human soul is a fragmentary portion of the divine substance.

Neither view (says Pascal) is sufficient in itself, but nor are they complementary. Pyrrhonism is not sufficient because it has so far forgotten the strength and goodness of human nature as to despair of any human ideal (571). Stoicism is not sufficient because, dazzled by memories of some earlier perfection, it has never come to terms with the miserable perversity of man's nature. What both philosophies ignore – one having forgotten it, the other never having learned it – is the gulf separating man as he is from man both as he was and as he ought to be. But because their emphases are so different, this is not to say that they meaningfully complement each other. Quite the reverse: they cancel each other out in bewilderment and confusion! 'One establishing certainty and the other doubt, one the greatness of man and the other his frailty, both destroy each other's truthfulness and falsity. So that, not being able to exist independently because of their shortcomings, and not being able to unite because of their contradictions, they crush and cancel each other out in order to make way for the truth of the Gospel' (572). Thus Pascal is led, so soon after his conversion, into a passionately eloquent eulogy of the Gospels, their extraordinary capacity for reconciling the seemingly irreconcilable appearances of life, their message of the twofold nature of man (the weakness exposed by Montaigne stemming from human nature, the grandeur admired by Epictetus inspired by supernatural grace). And Pascal concludes his meditation with the thought that these opposites and diversities are to be found perfectly conjoined and exemplified in the Person of Jesus, God and Man.

At several points in this tremendous discourse one senses Pascal's interlocutor – Le Maître de Saci who had been so reluctant to take on the new convert – now quite breathless with admiration. Here, it seemed to Saci, was a man who had come via profane philosophy to a religion of the Personality of Jesus indistinguishable from that of St Augustine and the other Fathers. And indeed the verve with which Pascal conducts his exposition of his faith is truly astonishing. The marshalling of detail, the building-up of a massive architectonic structure from items of miscellaneous reading are so sweeping and forceful that Saci clearly has great difficulty in keeping up. But he is the first to appreciate Pascal's masterly exposition of Montaigne, and to compliment him upon it (568). The charm and persuasiveness of Pascal's personality is well attested, even by the Solitaries themselves (561); in the *Conversation with M. de Saci*, however, the torrential eloquence is paramount: the rapid flow of ideas, the unerring directness, the pell-mell jostling of sparkling imagery! 'So he [Montaigne] rejects out of hand that Stoic virtue which is depicted with a stern face, a wild look in its eye, hair standing on end, with a wrinkled and sweaty forehead, in a tense awkward posture, far removed from men, in gloomy silence, alone on top of a rock: a phantom, he says, capable of scaring children, endlessly at work, only seeking rest but never finding it. His figure of virtue is artless, friendly, merry, charming and playful . . . ' (570) Even in translation, one can hear the insistent accent of the man. Pascal, hailed as a genius even before his arrival at Port-Royal des Champs, must have been recognized as much more of one by the time he left.

Already, in the Saci conversation, Pascal foreshadows the great outline of the *Thoughts*: its unique blend of philosophy, knowledge of the human heart, and veneration of the Christ figure. Also during these ten days at Port-Royal des Champs, Pascal seems to have found the time to compose a deeper meditation on the personality of Jesus, another element foreshadowing the *Thoughts* which later in fact was to be incorporated into it. *The Mystery of Jesus*, probably composed shortly after Pascal's conversation with Isaac de Saci but not published until 1844, is one of the supreme texts of mystical contemplation, trenchant in its lyricism, supple, poetic and hauntingly beautiful. It is a contemplation of the Passion of Our Lord, recalling the *Words on the Mystery of Jesus* written by Jacqueline three and a half years earlier. But whereas Jacqueline's meditation had concerned the actual death of Jesus, relating in each case some circumstance of His death to her own life which had become a death to the world, Pascal's on the other hand is dynamic rather than static. Infused with a strong (indeed Baroque) sense of dramatic movement, it is the record of an ongoing experience in the relationship of the world to the cosmic Jesus. 'Jesus', he writes (919*), 'will be in agony until the end of the world. There must be no sleeping during that time'.

The actual death on Calvary is but a tiny episode in the architecture of the *Mystery* as a whole. El Greco-like, Pascal rises to enormous heights of spiritual intensity in his picture of the Agony in the Garden of Gethsemane:

> Jesus seeks some comfort at least from his three dearest friends, and they are are asleep . . . And so Jesus was left to face God's wrath alone.

> . . . Only Heaven and he know . . .
> He suffers this anguish and abandonment in the horror of the night . . .
> I believe that this is the only occasion on which Jesus ever complained. But then he complained as though he could no longer contain his overflowing grief: *My soul is exceeding sorrowful, even unto death . . .* [6]
> Whilst Jesus remains in agony and cruellest distress, let us pray on.

Jesus, having died on the Cross (and the Cross is never shown in the meditation), ascends to His Father; and now begins the second movement of *The Mystery of Jesus*, with the impassioned dialogue between the eternal Christ, Risen, Ascended and Glorified, and Pascal himself:

> 'Take comfort; you would not seek me if you had not found me.
> 'I thought of you in my agony: I shed these drops of blood for you . . .
> 'I am present with you through my word in Scripture, my spirit in the Church, through inspiration, my power in my priests, my prayer among the faithful.
> 'Physicians will not heal you, for you will die in the end, but it is I who will heal you and make your body immortal.'

Only now does the *Mystery* approach the crux of the dialogue, the mystical exchange of words:

> JESUS: Repent then of your secret sins and the hidden evil of those you know.
> PASCAL: Lord, I give you all.
> JESUS: I love you more ardently than you have loved your foulness. *Ut immundus pro luto.*[7] May mine be the glory, not yours, worm and clay'.

Lord, I give you all: these are the only words spoken by Pascal in the actual dialogue with the Saviour. The entire accent of the *Mystery* is unselfishness. First and foremost is the unselfishness of the redemptive agony and death of Jesus, dying to save even those who would not keep watch with Him. In this first part of the spiritual drama Pascal the individual has no part to play. On the few occasions when persons other than Jesus are referred to, it is in the impersonal third person singular or the first person plural: 'We must tear ourselves away from those who are nearest and dearest to us . . . ', 'Let us pray on', 'We implore God's mercy, not so that He may leave us in peace with our vices, but so that He may deliver us from them'. Pascal does not think of himself in selfish or individual terms, but identifies with the whole worshipping community. The second and final part of the contemplative drama begins with a long homily from Jesus. Once again Pascal's personality does not intrude. His only words (five in all) are a complete oblation of himself. With the words: 'May mine be the glory . . . ' Jesus terminates the dialogue from the commanding heights of a *Maiestà*. All that remains is for Pascal – now, but only now – to speak at length in terms of himself. Even so, it is of himself, his weakness and inadequacies, in relation to the loving mercy of Jesus:

> I see the depths of my pride, curiosity, concupiscence . . . Jesus Christ the righteous . . . is more abominable than I, and, far from loathing me, feels honoured that I go to Him and help Him . . . Do small things as if they were great, because of the majesty of Christ, Who does them in us and lives our lives, and great things as if they were small and easy, because of His almighty power.

The Mystery of Jesus may be considered as a kind of rejoinder to the *Memorial*, a sustained reflection by Pascal on the mystical experience recorded on his parchment. Nowadays, indeed, both documents form part of the *Thoughts*. But the *Mystery* marks only the first stage in a studious and devout absorption in the life, death and transcendent love of Jesus which was to occupy him until his own death seven years later.

About this time Pascal may have drafted *On the Conversion of the Sinner* (548–52), of which he was probably the author although the attribution has not been definitively established. This manuscript found amongst Jacqueline's papers is undoubtedly very close to her brother both in style and content. It expresses the same revulsion from the world which she had seen in him and was to describe to her sister. According to this meditation, the soul 'touched' by God views both the world and itself in a new light; and such an experience fills the convert not with joy but with a fear inspired by the transience and nothingness of the worldly things which were once his greatest delight. He is no longer at ease in his enjoyment of worldly pleasures, yet even in his religious devotions feels sorrow and bitterness. To the soul of the new believer his body, mind, relatives, friends, enemies, wealth, poverty, disgrace, prosperity, honour, shame, respectability, contempt, health, sickness – the sky, the world, even his own life – count for nothing, since all except the human soul moves *into nothingness*. 'The soul, being immortal as it is, cannot find its happiness in perishable things . . . It begins to rise above the generality of men; it condemns their conduct, abhors their maxims, laments their blindness, and starts in quest of the true good' (549–50). This supreme object of human attainment has two characteristics: it is eternal, as is the soul, and incapable of being taken away from man without his consent; and it must be the most lovable, attractive and agreeable of all things. God is that supreme object of human attainment, the True Good; for the things of the world – though supremely attractive to the non-believer – can still be taken away from him against his will. No creature can be more amiable than his Creator, Whom the soul adores 'in silence, considering itself His vile and useless creature, and in repeated respectfulness adores and blesses . . . , and would wish to bless and adore . . . for ever' (551).

Probably later in 1655, but not at Port-Royal des Champs, Pascal transferred his attention from the cosmic and eternal Christ-figure to the defined (if ill-defined!) historical personality of the man Jesus, and drew up a detailed *Summary of the Life of Jesus Christ* under 354 headings. As its name implies, the *Summary of the Life of Jesus Christ* is not entirely unconcerned with the transcendent, eternal aspect of the Logos. The *Summary* opens with a Preface reminiscent of the Hellenistic Christology of the first Chapter of St John's Gospel. Its last three headings concern the general Resurrection of the Dead, the Second Coming, the Last Judgment, and the eternal life of the blessed. In the main, however, the *Summary of the Life of Jesus Christ* is – as it must be – primarily focused on the Gospel events, though Pascal also includes footnote references to the Old Testament,[8] the Pauline or quasi-Pauline Epistles,[9] the Jewish historian Josephus, the decretals of Pope Innocent III, the decrees of the Synod of Braga, and above all to the Fathers of the Church: Augustine, Ambrose, Jerome, Gregory, Tertullian, Leo, Cyril of Alexandria and Hilary of Poitiers. The very fact that there is such a wealth of Patristic reference indicates that the *Summary* could not have been composed at Port-Royal des Champs in January 1655. Besides, this text is such an intricate and eloquent conflation of many sources that, quite apart from the reading it presupposes, it would have taken weeks to produce. Based essentially on the Synoptic Gospels, so far as chronology is concerned, it attempts to harmonize the numerous contradictions between them and St John – such, for instance, as the fact that St John records three years of Jesus's ministry whereas the Synoptists record only one. Following St John (and, to some extent, St Luke), Pascal extends the ministry over three years. Similarly, even between the Synoptists themselves there are contradictions of detail. According to St Luke (XVIII 35), Jesus cures Bartimæus's blindness on His way into Jericho; according to St Matthew (XX 29) and St Mark (X 46), on His way out of that city; here, Pascal inclines towards St Luke – though his MS inaccurately refers to Jerusalem (640; **XI 47**). Again, according to Matthew (XXVIII 16), Mark (XVI 7) and also John (XXI 1), the risen Jesus appeared to His disciples in Galilee; but according to Luke (XXIV 13, 50), at Emmaus and Bethany in Judæa; Pascal skilfully if uncritically interweaves all these accounts (652, 654–5). So complete is his reconciling of the various discrepancies that no trace of them remains in the finished narrative. Frequently, especially about the time of the Crucifixion, Pascal adds his own commentary on the sacred events by way of exegesis. And at one moment in the *Summary* the tempo of narration perceptibly quickens: at the Agony in the Garden, when exegesis almost entirely vanishes, and the grim succession of events is evoked in staccato sentences or phrases, and when the reader is suddenly confronted with a detail not present in the *Mystery*. Just before Judas Iscariot approaches Gethsemane with the servants of Caiaphas, Pascal adds – much more pithily than Luke XXII 44 – that Jesus sweated blood.

Pascal's new-found interest both in the 'quest of the historical Jesus' and in the Church Fathers led him also, around 1655, to ponder the origins and growth of the Early Church. His thoughts on this subjects are set down in the *Comparison between Christians of Early Times and Those of Today*. The most noticeable difference he could see between the Church as it was in the early centuries of the Christian era and the Church in his own time was that in 1655, as against (say) 155, the sacrament of baptism was far more lightly regarded. In the earliest times (until, say, the fifth century and perhaps even as late as the ninth: though Pascal does not deal in dates) it had been a highly prized goal, only granted to adults or adolescents after a period of careful instruction. The result was, says Pascal (556), that the Christians of the early Church were all 'very well educated'. In his own day, however, he notes that the sequence of events has been reversed: the baptism of infants precedes the instruction of adolescents, and the worst aspect of this mockery of the sacrament of baptism is that the instruction may never be given at all. For the salvation of the souls of those who die young, the Church has consented to baptize babies (perhaps, the modern reader of Pascal may think, Pelagius was right after all?). Around him, in 1655, Pascal saw an ignorant and apathetic laity, for whom the word 'Christian' was nothing more than a conventional label attached by polite society. Uninstructed in their faith, lax in their performance of their religious duties, Christian people had allowed the world to interpenetrate with the Church – so corrupting it from the ideals established by its Founder. This antithesis between 'the Church' and 'the world', so characteristic of Jansenism, may serve to remind us that St Augustine had been a Manichean before becoming a Christian. But does it mean that Pascal was a Jansenist?

In the first place, it has not been proved beyond all doubt that the *Comparison* is by Pascal – though his authorship of it is extremely probable. More to the point is the fact that the *Comparison* is not in any sense a formulation of doctrine: it is more in the nature of a pastoral and spiritual tract, less grounded in history than typology. Rather as Rousseau does in his *Discourse on the Origin of Inequality amongst Men*, Pascal constructs an image of early times in order to point a lesson in the present. So high now are his religious standards and ideals that he sees, and is distressed by, the way in which the world falls far short of them. More than anything else, the *Comparison* embodies a moral judgment on his own time: not merely on his contemporaries' churchmanship but on their way of life; and it is a moral, rather than moralizing, judgment because it has all the warmth of private spiritual conviction. In all these ways the *Comparison* (if it is by Pascal, as the internal evidence suggests) foreshadows the *Provincial Letters*, with their emphasis on moral conduct rather than theological niceties, contrition rather than attrition, repentance arising from the love rather than the fear of God, and penance as an inner spiritual impulse rather than an external mechanical practice.

Pascal's earliest religious writings have, therefore, two predominant themes: an accent on the energizing morality of sincere and well instructed religious belief, and an intense devotion both to the mystical and the historical personality of Jesus. That Jesus has a historical personality is already a cornerstone of Pascal's religious outlook; and the *Summary of the Life of Jesus Christ*, for all its naive treatment of chronology, is a serious and sustained attempt to expound that life and personality in historical terms. The Father of Jesus is, as the *Memorial* emphasizes, the 'God of Abraham, God of Isaac, God of Jacob, not of the philosophers and learned men' (913*). The *Conversation with M. de Saci* transcends the clashing contradictions of secular philosophy to find a certainty beyond mere abstraction or subjective impression – a certainty typifying, and guaranteed by, moral experience – in the person of the 'Man-God' (572). The *Mystery of Jesus* is a supremely individual expression of belief in the cosmic and eternal power of Christ to seek out, redeem and uplift the sinner, and to inspire him with both the desire for repentance and the means to amend his ways. The *Comparison between Christians of Early Times and Those of Today* insists, however, that the sacrament of baptism (technically, the admission to Christ's Body and the fellowship of the Church) is no magical remedy for sin, automatically ensuring purity of life and high moral integrity: for 'the instruction which was necessary for the sacrament became voluntary, and then was neglected and eventually almost abandoned' (558).

It is easy for Bremond to claim that Pascal 'exalts the Mediator, but conceals and exiles God',[10] for Knox to complain of his 'Marcionism' ('the figure of the Redeemer so fills the canvas, as to obscure all thought of God in his eternal attributes'),[11] to scoff at the *Memorial* 'as if two Gods existed, and he, Pascal, were determined that his petition should go to the right address', and for the poet Valéry to remark that 'Pascal had found, no doubt because he had stopped looking'.[12] The core of Pascal's religious thought, even at the outset of his spiritual career, is that belief entails ceaseless vigilance, not supine self-satisfaction: the cosmic Christ, as distinct from the Jesus of Gethsemane, 'will be in agony until the end of the world. There must be no sleeping during that time'. His devotion to Christ is no diffuse, undisciplined *Schwärmerei* but an initial visionary experience supported and continually enriched by attentive study of the Gospels (hence the *Summary*): in St Paul's words, *I know Whom I have believed*.[13] To his way of thinking, the 'God of Jesus Christ' (554) is the only God who adequately responds to our human condition; and, whatever may be said about Pascal's somewhat quaintly termed Marcionism, Jesus Himself said: *No man cometh unto the Father, but by Me. If ye had known Me, ye should have known My Father also.*[14] Devotion to Jesus, a desire to live one's life in the shadow of His perfection, are in any case superior to the theological 'cavilling on the ninth part of a hair' which now threatened not only to destroy Jansenism but to rend the French Church with dissension.

Arnauld, at the time of Pascal's first visit to Port-Royal des Champs, was becoming increasingly embattled in the defence of Jansenism. In his judgment, the verdict of Pope Innocent X that the Five Propositions were to be found in the *Augustinus* made it all the more necessary to go on stressing the sharp distinction between *droit* and *fait*: the Pope may no doubt have been right to condemn the Propositions as heretical, but was he right – as a matter of literal fact – in asserting that the Five Propositions accurately summarized what Jansen had said? The situation became still more tense and confused when, in the autumn of 1654, the Duc de La Rocheguyon, Marquis de Liancourt was refused absolution at the church of Saint-Sulpice on suspicion of Jansenist heresy. Liancourt's chaplain was a Jansenist, and his granddaughter a nun at Port-Royal de Paris. Charles Picoté, the duke's Jesuit confessor, would not grant him the sacrament unless he first dismissed his chaplain and withdrew his granddaughter from her convent. Liancourt was also asked for assurances as to his own religious orthodoxy. Arnauld, feeling himself attacked, immediately set to work on a letter to his Jansenist friend the Duc de Luynes who not only made frequent retreats at Les Granges himself but who was increasingly to shelter Arnauld, Nicole and other Jansenists at his château of Vaumurier. *This Letter from a Doctor of the Sorbonne to a Person of Rank,* published on 24 February 1655, joined in the condemnation of the Five Propositions, arguing that the only true and reliable teaching in these matters came from St Augustine. To which François Annat, the Jesuit confessor of Louis XIV, replied in a letter accusing Arnauld of Calvinism. On 10 July, in his *Second Letter to a Duke and Peer* (again, Luynes), Arnauld returned to the attack. The fact was, he stressed, that the five controversial Propositions were not contained in the *Augustinus* at all. Only the first, by any stretch of the imagination, could be said to be there. The last four were not present in the text verbatim. And even the first Proposition – relating to the withholding of grace even from the elect – had been snatched unscrupulously from its context. After all, St Peter himself – the first Bishop of Rome – had been lacking in God's saving grace on the night of the Agony in Gethsemane, when he had denied his Master three times.

The distinction between *droit* and *fait*, which Arnauld maintained in both letters, seemed to the Jesuits a plain defiance of Papal authority. A commission of the theological faculty of the Sorbonne was set up, to try Arnauld's letters on both counts. But the writer of the letters had gone into hiding on 14 October and in any case was not allowed to defend himself. On 2 December, after two days' deliberations, the commission announced that it would defer its judgment. On 14 January 1656 Arnauld was condemned on the '*de facto* question' (namely, it was asserted that the Five Propositions did accurately represent Jansen's thought). However, the '*de jure* question' still remained to be decided. It was foreign to Arnauld's nature not to rise to a challenge. Whilst the suspense continued, a private meeting was held at Port-Royal des Champs to determine how best to react. According to the historian (and writer of fairy tales) Charles Perrault, the idea of widening and popularizing the debate – taking it out of the narrow, dry-as-dust theological faculty into the much wider and more generous public arena – was first mooted by one of his own brothers, after hearing the matter knowledgeably discussed in a family conversation: this brother mentioned his idea to the Duc de Luynes's steward at Vaumurier, Nicolas Vitart, and so it passed to Arnauld and his circle.[15]

Either by chance or design, Pascal was staying at Vaumurier – which, it seems, was Arnauld's hiding-place. This visit to Vaumurier was the second and last of Pascal's fully authenticated visits to Port-Royal des Champs. Marguerite Périer relates[16] that the first attempt at a more popular style of letter was made by Arnauld. But the heavy, clogging style of his two letters to the Duc de Luynes held out little hope that he would prove equal to the task. Having composed his letter, he read it aloud to the assembled Solitaries but, far from receiving an ovation, was met with a disapproving silence. Turning to Pascal, who was one of the audience, he is said to have exclaimed: 'But you are young! You should have a try!' And so Pascal, whose only previous literary experience in the public domain had been the writing of a paper on conic sections, an account of his *New Experiments Concerning Vacuums* and the *Narrative Account of the Great Experiment on the Equilibrium of Liquids*, now turned his hand to religious polemics. Within eighteen months of this secret meeting at Vaumurier, he was to produce eighteen bravura exercises in satire worthy of Demosthenes, Lucian, Juvenal or Swift: work so seemingly hostile to religion as to qualify him for inclusion in the *Index Librorum Prohibitorum* (**VII 229–32**; 6 September 1657).[17] The first of his *Provincial Letters* was rapidly drafted, read out to the Solitaries, admired, and brought to Perrault's brother in Paris by Vitart with the words: 'Here are the fruits of what you said to me a week ago'.[18] It was dated 23 January 1656.[19]

The series of eighteen letters – which could have gone on seemingly indefinitely except that by 24 March 1657 further publication seemed hopeless – are the nimble, quick-witted response to an ever-changing, and alas! constantly worsening, political and tactical situation. They are a running battle with the Jesuits, with each skirmish fought on a different territory, but with Pascal always dictating the action and never losing the upper hand. The irony was that, as each military encounter was won for the Jansenists by Pascal, so their political standing deteriorated. Triumphing through their brilliant young champion in the literary and ethical battle, they lost the war. For an *ethical* battle it soon became: Pascal quickly left the arid territory of the doctrine of grace, and moved on to the area of the moral consequences of religious belief which was both his own particular concern as a believer and also the area he felt, rightly, to be the Achilles heel of the Jesuits, the facet of their activities that was most damning. For all their skilful Biblical scholarship, philosophical study, schoolteaching, and missionary work in Africa, Latin America and above all in Ming and Manchu China, where, besides preaching the Christian Gospel for almost two centuries, they also gave valuable service as Court astronomers, mathematicians, painters and engineers, it is no exaggeration to say that the Jesuits have never recovered from the stigma attached to them by Pascal's *Letters*. For many people the word 'Jesuit' is almost a term of opprobrium. 'Casuistry', the resolution (in the confessional) of cases of conscience, and the subject of most of the eighteen *Letters*, has nowadays become practically synonymous in popular speech with 'devious, meaningless hair-splitting'. The philosophical propositions that 'might is right' and that 'the end justifies the means', which will for ever be associated with the Jesuits, owe this association to Pascal more than to anyone. Rightly or wrongly, Pascal in these *Letters* was to brand them with the indelible imputation of time-serving, worship of expediency, glorification of material power, reduction of religious observances to mechanical formulæ, and a fundamental amorality. The fact that the whole Society of Jesus was officially proscribed throughout Europe between 1773 and 1814, on the initiative of Pope Clement XIV, may or may not have been partly due to the gradually deepening and more pervasive impact of Pascal's *Letters*; it certainly underlines the justice of many of Pascal's criticisms.

Thus – and this is true of both Pascal's literary masterpieces – the *Provincial Letters* were not constructed according to any prearranged, organized plan. Events occurred during the writing of the letters between January 1656 and March 1657 which radically altered their tone, content and objectives. The first of these was the formal condemnation of Arnauld, on 31 January 1656, on the so-called *'de jure* question'. Having already decreed that the Five Propositions did correspond to Jansen's teaching, the Sorbonne now declared unequivocally that they were heretical. Arnauld was stripped of his doctorate of divinity and banished from the university. He lived quietly, in a kind of semi-exile, at Vaumurier, preserved from discovery and arrest by the social and political prestige of the Duc de Luynes. Undoubtedly the authorities must have suspected his whereabouts, but deemed it prudent not to press the matter; he had, after all, been silenced and was out of harm's way.

In March 1656 the persecution of the religious community at Port-Royal began. A criminal investigation was instituted both at the convent and at Les Granges for incriminating evidence of the fifth *Provincial Letter*, which had been published ten days before on March 20. Both by Pascal himself and those of the community who were in the know, the authorship of the *Provincial Letters* was a closely guarded secret. Eventually, in the hope of capturing their pseudonymous author, and silencing him as effectively as Arnauld had been silenced, the Solitaries of Port-Royal des Champs were arrested, threatened and scattered in December 1656. Yet, from its undiscoverable source, at intervals of roughly a month but sometimes oftener, the embarrassing indictment of the Jesuits still continued. In order to avoid detection, Pascal himself frequently had to go into hiding. One hiding-place was an inn, 'At the Sign of King David', in the Rue des Poirées just by the Sorbonne and opposite the Jesuits' Collège de Clermont, an inn where his brother-in-law Florin Périer was also in the habit of lodging. He sought refuge, too, with the Duc de Roannez in the Cloître Saint-Merri in June 1656. Pascal, of course, was not being actively pursued by the authorities, as he was not officially the author of the *Provincial Letters*; he was lying low. In late November and early December 1656 he may also have put in a visit to Vaumurier.

Arnauld meanwhile moved secretly to and fro between Vaumurier and Paris. His closest assistant amongst the Solitaries, Pierre Nicole, was near but not with him. Between them, they supplied the ammunition which Pascal fired with such devastating effect. Arnauld lived for a time with a M. Hamelin in the Faubourg Saint-Jacques – very close, in other words, to Port-Royal de Paris. Nicole's place of concealment was the Hôtel des Ursins, on the Ile de la Cité – within easy reach of Pascal when the latter was at the Hôtel de Roannez. On occasions, in April 1656 for instance,[20] Pascal and Arnauld were able to make contact more or less every day. In September and October 1656 Pascal, Arnauld and Nicole all appear to have been in Paris at the same time.

They were also probably all in Paris in August of that year, at the time of the tenth and eleventh *Letters* and of the reply which Jacques Nouet was stung into publishing towards the end of the month. In his *Replies to the Provincial Letters Published by the Secretary of Port-Royal against the Fathers of the Society of Jesus, Concerning the Moral Teachings of the said Fathers* Jacques Nouet, a Jesuit priest, fiercely denounced Pascal for his textual misquotations, particularly stressing the unknown author's alleged misrepresentation of the Jesuits' views on almsgiving, simony and bankruptcy, in *Provincial Letters* VI and VIII. Nouet was later to acquire a justified reputation as a devotional writer of great spiritual insight, and to reply to the *Provincial Letters* in such narrowly captious terms was unworthy both of his own outstanding gifts and of the many visible merits of his religious society – on which none of Pascal's detractors seems ever to have thought of resting his case!

All this was merely the cut and thrust of the debate which Pascal, Arnauld and Nicole had freely entered into on their own terms, and where anyway they were the masters of the argument. They had nothing to fear from Nouet's feeble protests and tired rhetoric! Within two months of the appearance of this pamphlet, they and their cause had been dealt a much more formidable blow. On 16 October 1656 a further condemnation of the *Augustinus* was promulgated by Pope Alexander VII. To this Papal constitution *Ad Sacram Beati Petri Sedem* [21] was attached a Formulary of Submission[22] requiring that all priests, monks and nuns of the French Church should sign it, thus accepting the condemnation of the Five Propositions and effectively dissociating themselves from Jansenism. The penalty for failure to comply with this requirement was excommunication.

A further indignant riposte was voiced by the Jesuits in December 1656 when the King's confessor, Annat, published *The Good Faith of the Jansenists in their Quotations from Authors, as seen in the Letters which the Secretary of Port-Royal has been circulating since Easter.* Not for the first time, nor the last, the serious accusation was levelled against Pascal and his friends that they had been unscrupulous in their quotations from Jesuit authors: wrenching them out of context, misrepresenting them or even actually changing the words. Both sides, therefore, accused the other of travestying the meaning of originals. If indeed Annat's allegations were true, this would only of course mean that Pascal was doing to the Jesuit authors what he and his Jansenist friends complained of with regard to the *Augustinus.* For a man who so adamantly denied that the end could ever justify the means, it was an unacceptable allegation; but one, nevertheless, that he could deal with and did defend himself against in the eighteenth *Letter.* But once again a much more formidable blow was dealt to his cause by a power which, in the short term at least, was much mightier and deadlier than the pen.

On 11 March 1657 Alexander VII's Bull *Ad Sacram Beati Petri Sedem* was formally presented by the Papal Nuncio to Louis XIV, and the decision to impose the Formulary of Submission was taken by the Assembly of the French Clergy six days later. Now that Arnauld had been condemned, disgraced and expelled from the Sorbonne, and with all priests and regulars compelled to acquiesce in the outlawing of Jansenism, there was no further reason for Pascal to continue his dazzling defence of a complexion of religious faith which perhaps was not entirely his. He had abundantly made his point about the Jesuits' moral instability. And so the *Letters* tailed off to what – in worldly terms – was a truly ignominious end, with just a fragment of a Nineteenth Letter (902–4) surviving which remained unpublished for over a hundred years.[23]

But another event had occurred during the writing of the *Letters* which was of incalculable importance both to Pascal himself and, more particularly, to the future conception of the *Thoughts.* On 24 March 1656, four days after the publication of the fifth *Letter,* a miracle – or what Pascal, his family and friends held to be a miracle – occurred at the convent of Port-Royal de Paris. The beneficiary of the miracle was his ten-year-old niece and goddaughter Marguerite Périer, the third child of his sister Gilberte. Marguerite had been suffering for three years from a fistula of the eye, a peculiarly painful and unsightly affliction – long, thin and ulcer-like – which in that particular eye had induced semi-blindness. Various treatments had been prescribed for it, but all to no avail. It seemed as if she would have to resign herself to the disease for the remainder of her life.

A thorn from the Crown of Thorns, traditionally worn by Jesus as He stumbled to His Crucifixion and hung on the Cross, had recently been presented to Port-Royal by a priest who was a relation of Mère Angélique, the Abbé Le Roi de La Poterie. It was to be placed, as an object of veneration, in a reliquary in the chapel. The day chosen for its solemn installation was the eve of Lady Day. All the pupils of the girls' school at Port-Royal de Paris, including Marguerite, were to walk past it in procession, making their solemn obeisance. As she approached, she was told by the nun in charge of the procession to pray for her recovery. The nun placed the sacred relic against her eye. The pain that had tormented her for three years rapidly subsided. Within a few hours of the application of the sacred thorn to her eye, all the swelling had disappeared. And a few days later she was as healthy as if the fistula had never been.

At first, this 'miracle' having occurred in a Jansenist convent, it too was the object of virulent attacks. For the cure of Marguerite Périer's disease there may or may not have been an adequate medical explanation. Sainte-Beuve, for instance, who had some scientific knowledge, believed that 'little Marguerite did not exactly have a fistula but a lachrymal *tumour* caused by the obstruction of the lachrymal canal . . . This obstruction was evidently *partial*, since, if the tumour was pressed, some of its contents oozed out through the lower orifice of the canal, as indeed should be the case'.[24] In the cooler retrospect of modern medical knowledge, it may well be (as Sainte-Beuve suspected) that the healing of the disorder was due to the external application of the metal reliquary: pressed, he suggests, against the lachrymal canal with all the energy of religious fervour. But to Pascal, as to everyone in the Jansenist circle, this cure was clear evidence of divine intervention and supernatural grace; no doctor called in by the convent could account for the almost instantaneous recovery. From September, or perhaps as early as August, 1656 Pascal was busily engaged on a long letter or treatise which would be a justification of miracles.

The existence or otherwise of miracles has always been a contentious matter not only in the frequent historic disputes between science and religion but even within the Church itself. Malebranche, a follower of Descartes and an Oratorian priest, held[25] that, if a miracle were actually to occur, it must be as a consequence either of *unknown* general laws or of *contingent* circumstances foreseen by the divine mind throughout eternity. The Godhead, even according to Malebranche, can exist only in a perfectly ordered universe from which the disorder and illogicality of miracles is banished. For not even God can make two and two equal five: to do so would be to contradict His own essence, to destroy the rationality of a system whose ultimate Reflection and Principle He is.

Certainly from the time of the cure by the Sacred Thorn, Pascal, however, considered the matter quite otherwise. He could not see how God, being by His very nature omnipotent, could preclude Himself from the possibility of momentarily disrupting the laws and harmonies of the universe. To God, after all, everything is potential as well as actual. The Bible itself – both Old and New Testaments – is full of miracles: God speaking from the burning bush, the crossing of the Red Sea, the falling of manna from Heaven, the raising of Lazarus, the changing of water into wine, the feeding of the Five Thousand. The very contingency of the created world is of itself a miracle, as is the existence of Jesus Christ perfect God and perfect Man, dead yet alive, temporal yet eternal. God, by His nature being mysteriously twofold, moves and has His being both within the world of the unchangeable Absolute and within the realm of the finite and contingent. Malebranche's views are in this, as in some other respects, a simplistic anticipation of eighteenth-century deism. Pascal, on the other hand, places the Christ figure firmly and devoutly at the heart of Christianity. The importance of Marguerite Périer's sudden and remarkable restoration to health lies less in the encouragement it gave to his jaded spirits at the height of an acrimonious controversy than in the emphasis which it induced him to lay on miracles: its illumination of the role of the miraculous in the world, not only through the power of spiritual healing but (ultimately) through Jesus's very self-revelation.

Even now, it seems, Pascal may have begun to put together notes towards his massive justification of the Christian religion, the *Thoughts*, in which the importance of miracles as proof of Jesus's divinity is seen as paramount.[26] But as a more tangible and immediate expression of his new-found realization of the miraculous he adopted, as his personal device for engraving on a signet-ring, a Crown of Thorns in glory, borne aloft by clouds against a background of sun-rays,[27] both of which are customary heraldic symbols of the Divinity. Beneath the Crown, and uniting the miraculous experience of March 1656 with the miraculous and even more personal experience of November 1654, was the motto *Scio Cui Credidi*: 'I know Whom I have believed'.

During the writing of the *Provincial Letters* Pascal was particularly close to the Roannez family, the twenty-nine-year-old duke and his twenty-three-year-old sister Charlotte, both of them unmarried. For some time, even before the fierce quarrel with the Jesuits, Roannez had allowed him the use of a room in his own house in the Cloître Saint-Merri. Pascal could come and go as he pleased between his house and the duke's. Thus, when the storm broke, his refuge in the Cloître Saint-Merri was a ready-made one, where he enjoyed in some measure the privilege of ducal protection. In the spring of 1655, under the influence of Pascal's new devoutness and of Singlin with whom Pascal had put him in touch, Roannez had been converted – if not to Jansenism, then certainly to a much deeper Christian piety and a desire to have done with the world. He decided that, as soon as he could obtain the King's permission to do so, he would sell his governorship of Poitou and retire to live with the Oratorian Fathers.

Some years previously, around 1651, he had had thoughts of marriage. The girl on whom he had set his heart was Antoinette-Louise de Mesmes, 'the richest heiress in the kingdom' (39), who was not yet of marriageable age. Roannez 'never even doubted that he could have her, because at that time he was the only unmarried duke and peer, for in those days there were very few dukes.'[28] Perhaps Roannez's interest in her had been noted by her family: the fact was that, when the time came for her to marry, they themselves suggested the marriage to his great-uncle the Comte d'Harcourt. It would no doubt be untrue to suggest that Pascal had actively dissuaded his friend from marrying; even so, the desire – indirectly inspired by Pascal's example – to abandon the world and all its snares had brought about a complete change of mind in Roannez. To his uncle's incredulous dismay, he was no longer sure that he wanted to marry Mlle de Mesmes; he asked for time in which to think the matter over (and perhaps to consult Pascal?). A fortnight later he pronounced against the marriage: worse still, from the uncle's point of view, he declared himself unwilling to marry at all. Harcourt flew into a rage, and the main target of his anger (whether justifiably or not) was Pascal. His animosity spread like wildfire through the Hôtel de Roannez, and against Pascal who at that time was in residence there. About eight o'clock one morning the concierge of the duke's town house crept upstairs to Pascal's bedroom, intending to murder him with a dagger. But, contrary to his usual habits, Pascal – it seems (39) – had gone out early into town. And so, providentially, he escaped death or injury and moved away, at least for a time, from the Hôtel de Roannez. Until his death in 1696, at the age of sixty-nine, Roannez persisted in his refusal to marry: finally selling his governorship of Poitou to the Duc de La Vieuville, he took to wearing a monastic habit, although he never made his profession; he withdrew to a religious house. Four years after the upset of her marriage plans, Antoinette-Louise de Mesmes married the Comte de Vivonne.

Charlotte de Roannez, the duke's sister, showed an almost equal aversion from marriage. At first, admittedly, she was attracted by the prospect of having a husband and an establishment of her own. The Marquis d'Alluye et de Soudis was tenderly attached to her. In the summer of 1656, having developed an eye disorder, she was persuaded to follow Marguerite Périer's example and make a novena before the Sacred Thorn at Port-Royal de Paris. These nine days were sufficient to fill her with a sense of holiness.

Pascal's nine surviving letters to Charlotte and her brother Artus de Roannez (505–18) are in fact principally addressed to her. Written between September and December 1656, they belong to the period when Charlotte, after her novena at Port-Royal de Paris, had become aware of her religious vocation. Like Jacqueline Pascal six years before her, she was still living at home but felt irresistibly drawn to the monastic life. Her mother the Marquise de Boisy was unhappy about the direction which Charlotte's life seemed to be taking. Several times Pascal alludes to the fact that he was acting as an intermediary between her and her secret spiritual director Singlin. His letters, therefore, served both as consolation and encouragement.

Her changed character since the novena has, he assures her (511), produced an entirely unfamiliar accent in the language of her letters to him:[29] this 'new heart',[30] he says, has found expression in a new language. But in his letters to her we may also detect a new language. It is not merely their subject-matter – the Hidden God [31] (510), the unity of the Church expressed in devotion to the very Pope who was condemning Jansenism at that time (513–14), the absolute necessity of living in the present (517), the eternal restlessness on earth even of those who are to be saved (507–8), the fact that all earthly things are but the symbol of an eternal

Reality (515) – which is so startlingly novel in many respects, foreshadowing the *Thoughts*; the language in which Pascal's ideas are couched has also taken on a new mellifluence, a new mansuetude. 'It is very certain', he remarks, referring to Charlotte's grief at her enforced separation from Port-Royal (507),

> that you never detach yourself from things without pain. You do not feel you are chained when you voluntarily follow that which is leading you on . . . But when you start to resist as you walk away, then you really suffer; the chain stretches and suffers all the violence; and that chain is our own body which does not break until death.

> True piety . . . is perfect only in Heaven . . . If there is any sadness intermingled . . . , it comes from ourselves and not from virtue; for it is not the result of the piety which is stirring within us, but of the impiety that still remains (516).

> For sixteen hundred years [the Church] has been in travail for you. It is time to be in travail for her, and for all of us together, and to give her all that remains of our lives, since Jesus Christ only took His in order to lose it for her and for us (514).

> The things of this world, however new they may be, grow older, the longer they last, whereas this new spirit renews itself more and more, the longer it lasts (512).

And the *Letters* also contain wonderfully sustained passages of sober eloquence: images like the ingeniously wrought parallel between the spurned God and the exiled Prince (509), a Baroque conceit worthy of Donne; the emphatic assertion that Christianity is a religion not of sadness but of pure joy (515); and the almost mystical justification of Catholicism in terms of the God Who is invisible and yet visible (510):

> *Verily Thou art a God that hidest Thyself* . . . [32] The veil of nature which covers God has been penetrated by several unbelievers, who, as St Paul says,[33] have clearly seen an invisible God in the visible created world. Heretical Christians [e.g., Protestants] have known Him through His humanity and worship Jesus Christ as God and man. But as for recognizing Him beneath the host, that is

> the special attribute of Catholics alone: only we have been enlightened by God to that extent.

The *Letters* testify also to Pascal's vast and growing reading of the Scriptures and Church Fathers (Paul, Isaiah, Ezekiel, Job, the Psalms, Augustine, Tertullian, Leo) and to his marvellous facility in pondering and turning to good account the fortuitous readings of the daily office: the epistles of the Seventeenth and Twenty-Third Sundays after Pentecost, 24 September and 5 November 1656 (508, 513), or the Vesper reading for Christmas Eve 1656 from Isaiah XXXV (518). Already the observances of the Christian religion were becoming an intensely personal preoccupation, an integral part of the routine and the meditation of his daily life.

5

The *Provincial Letters*

'If the *Provincial Letters* were serious, nobody would read them any more', Gide has written.[1] The *Letters* are in fact profoundly – even, at times, desperately – serious, but Pascal does not become pompously solemn or tediously earnest: he is never boring. Yet to many, if not most, people the subjects he is basically canvassing could rapidly induce boredom! In the first three, or even four, out of eighteen letters, he is concerned with the question of divine grace, a very intangible and metaphysical concept. Is God's grace freely given to all, as the Pelagians and those semi-Pelagians, the Jesuits, maintain? or is it restricted to the Elect? To be more theological still, is it *sufficient* or is it *efficacious*? Is it sufficient to enable a man to 'work out his salvation with diligence' whilst not ensuring that he will do so? Or does the very bestowal of divine grace ensure that the recipient will live virtuously and attain salvation? To chop logic about niceties which, if knowable at all, can be known only to the mind of God may to the modern reader appear to be verging on the ludicrous. Pascal makes such a potentially sterile discussion intensely fruitful and human.

The bulk of the *Provincial Letters* are, however, concerned not with the dogmatic theology of grace but with the moral theology of casuistry.[2] Casuistry, the application of ethical rules to cases of conscience, was another field in which the Jesuits had specialized to the point of making it peculiarly their own. Indeed, we owe it to the Jesuits – or perhaps to Pascal's somewhat one-sided picture of them? – that in certain quarters the word *casuistry* has become a term of abuse. To many, casuistry has come to mean sophisticated hair-splitting, specious special pleading; yet casuistry is in essence a very respectable and necessary department of moral theology, not invented by the Jesuits but with a distinguished ancestry extending back into the Middle Ages: necessary because wherever the confession of sin is made, not directly to God, but indirectly through the intermediation of a priest, some form of guidance must be available to the priest which will enable him to instruct and direct his penitent.

Well before the advent of Protestantism, a voluminous literature had arisen on every aspect of Christian ethics; and this was considerably added to, during the Counter-Reformation period, by Jesuits (mostly Spanish) such as Luis Molina, Gregorio de Valencia, Francisco Suárez, Gabriel Vasquez, Antonio de Escobar and Leonard Lessius. Just as Molina had been the proponent of a modern semi-Pelagianism, so Escobar diluted the rigour and astringency of the Church's moral teaching. Both men, but Escobar in particular, sought to make the Catholic faith more acceptable to those who were in real danger of falling into Protestantism.

It is a commonplace of moral theology that no two cases of conscience are ever exactly alike. Faced with a variety of moral judgments, the confessor may (in ascending order of inflexibility) be a *Probabilist*, a *Probabiliorist* or a *Tutiorist*: he may, in other words, adopt a consistently lenient attitude, or else the attitude which in all the particular circumstances of the case he considers to be the fair and right one, or else he may adopt a consistently hard line. The Probabilist will always seek the most lenient judgment, even when it is less likely to be the safe, or correct, one. The Probabiliorist will seek the most lenient judgment *only when*, in his opinion, it is more likely than the less lenient judgment to be the safe, or correct, one. The Tutiorist will always seek the least lenient judgment, even when it is less likely to be the safe, or correct, one than some more charitable interpretation of the facts. But not even Escobar denied that the Probabilist must have some solid ground for the moral judgment he proposes, even though that judgment is less likely than other types of judgment to be just and correct.

Pascal accepts the necessity of casuistry, but insists that in the hands of the Jesuits it has become a depraved and distorted thing. Of all the Jesuit practices he abhors, Probabilism is the one which incurs his keenest censure in the *Provincial Letters*. Cutting through the semantic entanglements of Tutiorism, Probabilism and Probabiliorism, he points out again and again that these are mechanical ways of viewing human sin and human destiny, formulæ which permit those who practise them to evade the true duty of the confessional which is to uplift and correct. All this, however, is done in the lightest and most fanciful way: by revealing and emphasizing the human aspect of every issue, by stressing that behind the theological subtleties lie carefully calculated, even devious ulterior motives, and by hinting – even in most jocular vein – that the Jesuits, far from 'simplifying', 'rationalizing' and 'modernizing' Christian theology, have in fact degraded and debased the Church's view of humanity. From this it is only a step to proving that they will in fact also, in the long run, debase the public's view of the Church.

How then are the distinctly human aspects of these two problems (grace and casuistry) brought into prominence? The first device is Pascal's invention of the Provincial Friend. His attack on the Jesuits is couched in the form of letters supposedly written by an intelligent gentleman – an *honnête homme* – to his friend in the country: the friend, too, is an intelligent man, but both are unversed in theology. The writer of the letters endeavours to explain to his country friend, in layman's language, what all the impassioned controversy is about. This approach enables Pascal to show up the fundamental unreality of the doctrinal dispute and, worse still, the immorality masked by an appearance of sweet reasonableness and forgiving leniency which lies at the heart of the Jesuits' ethical system. The endless charge and counter-charge of the debate about grace emerges, therefore, as shadow-boxing. The dispute about casuistry is proved to be no empty academic disputation about 'isms': for the very integrity of man both as moral agent and worshipping being is shown to be endangered by the Jesuits' debasement of the language in which the Church speaks to the world – their attempt to present God's yoke as easy and His burden light, even at the expense of encouraging man's hypocrisy and callousness. Pascal, by addressing his readers in the language of an intelligent layman, and by showing them the practical consequences in human terms of the Jesuits' new-fangled approach to moral theology (all this through the device of letters to an intelligent but mystified *provincial* reader), did more than anyone in his century – and perhaps since – to undermine the Jesuits' growing ascendancy.

Strange to say, no one before Pascal's time had thought of satirizing an opponent through the device of intelligent letters written by an apparently unbiased observer to dispel a friend's naive bewilderment. It was a method which was to have its imitators: Montesquieu, in his *Persian Letters*, a satire of French life, religious thought and ways of government supposedly written by two Persian visitors to Paris; and Voltaire's *Philosophical Letters*, exposing the weaknesses of French society by praising the virtues of the English. Voltaire indeed realized that the essence of Pascal's genius, in this respect, was not so much to have used the epistolary form, until then very largely confined to the novel, as to have perfected the use of naive irony. The device of the naive observer was to be used by him, with a success equal to Pascal's if not greater, in such short stories as *Zadig*, *Micromégas* and *Candide*. The very name 'Candide' epitomizes the apparently frank, impartial, sincere outlook on the world pioneered (as a device of exquisite irony) in the *Provincial Letters*.

The second device employed by Pascal in the *Provincial Letters* is that of the enemy (in this case, a Jesuit priest) damaging himself by the absurdity of his own remarks. Pascal's supreme skill is to present this Jesuit priest as 'a man more sinn'd against than sinning': a basically kind and well-intentioned man, friendly and likable, but unintelligent, simplistic and wholly misguided in his intellectual outlook. As a target for ridicule, and an example of the Jesuits' wicked folly, it would have been totally unconvincing for Pascal to have presented him, in lurid Mephistophelean terms, as an astute Satanic figure. As he is, we love him rather than hate him, pity him rather than scorn him, and look on him as a poor misguided fool unworthy, in his simplicity, of the devious double-dealing Jesuits but of whom, in its calculating heartlessness, the Society of Jesus is equally unworthy. His simplicity is such that he is invariably admitting what ought to be concealed, and not only making the admission but glorying in it – much as a commercial representative might sing the praises of his firm's latest invention: for to the Jesuits (Pascal insinuates it again and again) everything has become mechanical.

A fine example of his ingenuousness occurs in the discussion of Probabilism, in Letter VI, where Pascal is in the process of demonstrating that the Jesuits' new moral theology will permit or condone any crime or sin, however heinous. The priest artlessly observes that, according to the decisions of three successive Popes,[3] bishops who also happen to be regulars are not exempted by their worldly status as bishops from their monastic vow of abstinence from meat throughout their lives. Nevertheless, he continues, Antonino Diana (a Theatine father so renowned as a casuist that he was appointed to be the examiner of bishops) maintains that they are exempted from that vow.

> 'And how does he reconcile that?' I asked him.
>
> 'By the subtlest of all the new methods', replied the Father, 'and by the utmost refinement of probability. I will explain. As you saw the other day [707–12; *78–85*], the fact is that both the affirmative and the negative of most opinions have some probability, in the view of

our doctors, and enough to be followed with a clear conscience. This does not mean that the pro and the con are both right in the same sense – that would be impossible – but just that both are probable and consequently safe.

'On this principle our good friend Diana speaks thus in part V, treatise xiii, resolution 39: *I reply to the decision of these three Popes, which runs counter to my own opinion, that they have spoken in that way by adhering to the affirmative, which is indeed probable, even in my own view; but it does not follow from this that the negative does not also have some degree of probability.* And in the same treatise, resolution 65, on another subject where he also disagrees with a pope,[4] he speaks as follows: *That the Pope said this as head of the Church, I freely admit. But he did so only within the extent of the sphere of probability of his own opinion.* So you can see now that this is not offensive to the Popes' feelings; that would never be tolerated in Rome, where Diana is held in such high esteem. For he does not say that what the Popes have decided is not probable, but, whilst leaving their opinion within its full sphere of probability, he nevertheless says that the contrary is probable also.'

'How very respectful of him', I said.

'And it is subtler', he added, 'than what Father Bauny[5] replied when his books had been censured in Rome. For, writing against Monsieur Hallier,[6] who at that time was furiously persecuting him, he let slip the phrase: *What has the censure meted out by Rome got to do with that imposed in France?* You can see clearly enough from this that, whether by interpreting terms [715–16; *88–9*] ordetecting favourable circumstances [716–17; *89–90*] or, last but not least, by means of the double probability of pro and con [707–12; *78–85*], these alleged contradictions which previously astounded you can always be reconciled without ever offending the decisions of Scripture, Councils or Popes – as you can see!'

'O reverend Father', I replied, 'how lucky the world is to be governed by you! How useful these probabilities are! I did not know why you had gone to such lengths to establish the fact that one doctor, *if he is a serious doctor*, can make an opinion probable;[7] that the contrary may also be probable; and that people can then choose between pro and con just as the spirit moves them, even if they do not believe it to be true, and with such a clear conscience that any confessor refusing to grant absolution on the strength of these casuists would be in a state of damnation. From which I now realize that a single casuist can lay down new moral rules as he

> pleases, and decide in any way he thinks fit any matter of moral behaviour' (717–19; *91–2*).

In these words Pascal shatters the theological, and philosophical, basis of Jesuit casuistry. He mocks, first of all, the semantic haze with which the Jesuits had managed to surround the word 'probable', the lay meaning of which is: 'to be expected.' In the language of mathematics, if a bag contains x green balls and y white ones, and if except for the numerical difference between x and y we are as likely to draw green as we are to draw white, then the probability of drawing green is $\frac{7}{10}$ whilst that of drawing white is $\frac{3}{10}$. Turning from the language of mathematics to that of every day, we should say that out of a bag of seven green and three white balls *it is probable that* a green one will be drawn: it is merely *possible* (in ordinary parlance) that white will be drawn, whereas the same outcome in mathematical terms is *probable* to the extent that there is a $\frac{3}{10}$ probability – although it is, of course, *more probable* that the colour of the ball produced will be green!

As applied by the Jesuits to the moral teaching of the Church, such a quasi-mathematical use of the language of probability was bound to have far-reaching and, in Pascal's view, disastrous consequences. The foundation of the new casuistry was that, although a course of action recommended by the Scriptures, Councils and Popes was admittedly more *probable,* even the opposite course – if backed up by the authority of at least one recognized doctor of the Church – was also probable. By asserting that any course of action, even if condemned by a hundred ecclesiastical authorities, was still morally permissible if supported by one, the Jesuits had opened the way for a drastic easing of moral standards. They, after all, could supply on any issue the one minority opinion which made all things probable.

This general moral free-for-all, so vehemently denounced by Pascal in most of the *Provincial Letters,* has been considered by many commentators (not least by the Jesuits themselves!) to be a travesty of the truth. The *Provincial Letters,* writes Edwyn Bevan,[8] 'were a witty caricature which has had enormous influence in creating the popular idea of the Jesuits.' A caricature is a distortion of the truth in order to reinforce some salient aspect of it. Pascal's attack on the Jesuits' moral theology is not a caricature in the sense that it is either untrue or unfair. He is much too reasonable a logician to have it supposed that, because the *possibility* of an abuse of authority exists, that abuse will necessarily occur. The salient aspect of the truth about their position, in his judgment, is that the mere *ipse dixit* of one of their own number – however comparatively obscure – can of itself destroy the whole traditional teaching of the Bible and the Church throughout the ages. What he denounces is not so much the actual doings of the Jesuits taken as a whole as their *potentially* pernicious influence and, above all, the incoherence of their logical position. 'Behold him who taketh away the sins of the world',[9] François Hallier is supposed to have said of Étienne Bauny; and Pascal, in his Fourth Letter (694; 63), makes use of this elegant witticism in devastating mockery of his opponents.

Essentially, therefore, it is the unscrupulousness of the Jesuits which irks Pascal, an unscrupulousness which he exposes through the Jesuit father's naivety. Whenever he engages in conversation with the *honnête homme*, the priest finds himself impaled on the horns of a dilemma, or embroiled in a *reductio ad absurdum*. For it is plainly absurd that 'the double probability of pro and con' can exist on any moral issue. Not only is the priest hopelessly incapable of dealing with his interlocutor's irony, he is himself imprisoned within his creator's. Throughout his remarks runs a pathetic strain of complacent modernism, the belief that newer is better and that, in a never-ending march of progress, notable improvements are afoot within the Church. Hence Bauny's reference to 'the censure meted out by Rome.' It had seemed sufficient, and clever enough, to the Jesuits in 1641 or thereabouts to shrug off Papal condemnation with the Gallican remark that what really mattered was the approval of Paris. By 1656, however, a new refinement had been introduced into the system! Fifteen years after Bauny's slighting comment, it is now claimed that the opinions of Rome and Paris matter equally but that equally the teaching of one 'serious doctor' of theology in Paris is just as authoritative as the Pope's!

This unholy chaos in which anything seems ultimately permissible and all views equally probable is further exposed in a subsequent discussion between the *honnête homme* and the Jesuit father. In this conversation the artless priest is extolling the virtues of the new-found Jesuit doctrine of direction of intention. This doctrine is, of course, virtuous because it is so eminently useful.

> 'Well then, you should know that this marvellous principle is our great method of *directing the intention*, which is of such importance in our moral system that I might almost venture to compare it to

the doctrine of probability. You have seen some of its features in passing, in certain maxims I have outlined. For when I explained to you how footmen can run certain tricky errands with a clear conscience [724–5; *98–9*], did you not notice that that was only by deflecting their intention from the evil they are procuring and applying it to the profit which it brings them? That is what *directing the intention* means. Likewise, you saw that those who give money for benefices would be downright simoniacs if they did not also deflect their intention in that way [720–1; *94*]. But I now want to show you this great method in all its glory, in the matter of homicide, which it justifies in innumerable circumstances, so that you may judge from this all the results it is capable of producing.'

'I can already see', I said, 'that this will make everything permissible; nothing will escape.'

'You are always going from one extreme to the other', the Father replied. 'You must stop doing that. As evidence that we do not permit everything, note, for instance, that we never allow anyone to have the formal intention of sinning just for sinning's sake; and that if anyone insists on having no other end in evil-doing than evil-doing itself, we will have nothing more to do with him; that is diabolical; and to that we make no exception, whether of age, sex or rank. But whenever people are not in that unfortunate frame of mind, then we try to put into practice our method of *directing the intention*, which consists of setting up some lawful objective as the purpose of their actions. Not that we refrain from deterring men from forbidden things as far as is within our power; but whenever we cannot prevent the action, at least we purify the intention; and thus we correct the viciousness of the means by the purity of the end.

'This is how our Fathers have found a way of permitting the acts of violence involved in defending one's honour. For all you have to do is to deflect your intention from the desire for revenge, which is criminal, and apply it instead to the desire to defend your honour, which is permissible according to our Fathers. And that is how they fulfil all their duties towards both God and man. For they please the world by permitting such actions; and they satisfy the Gospel by purifying intentions. This is something the Ancients knew nothing about; this is something you must thank our Fathers for. Now do you understand?'

> 'Very well', I said. 'You allow men to operate in the external and physical realm of action, and you assign to God the internal, spiritual impulse of intention; and by means of this equitable allocation you unite human and divine laws . . .' (728–9; *103–4*)

Once again Pascal does not castigate the moral attitudes of all Jesuits at all times; he exposes the potential perniciousness of their ethical system. This is shown in its unmistakable colours as a system of expediency, ruthless opportunism and cynical disregard for others. It includes the doctrine that the end justifies the means; and the Jesuit father, in his stupid candour, actually goes so far as to use the words 'end' and 'means' in this extract. In Letter VI Pascal had already applied the same argument to the purchase of benefices, thus (in the Jesuits' view) exonerating such purchasers from the charge of simony. He had gone on, in the same Letter, to show how the Jesuits' doctrine taken to its logical conclusion will permit servants to obey the orders of their dissolute employers, carrying letters and presents, opening doors and windows, helping their masters to climb up to windows,[10] all with a clear conscience. The ultimate conclusion of their doctrine, however, is – as will be shown virtually throughout Letter VII – that not even human life is sacred. Homicide is justifiable 'in innumerable circumstances', and one of the most serviceable of their ethical discoveries is that all, even the taking of life, is permissible in the defence of one's personal reputation and integrity. No moral problem, says the Father, can arise provided one only 'sins' with a laudable purpose in view. Then sin is not sin; yet (in a touch of supreme irony) not even the Society of Jesus can condone sinning for sinning's sake. Even the Jesuits consider it 'diabolical' that anyone should actually insist on having 'no other end in evil-doing than evil-doing itself'! It is all the more diabolical that anyone should be so insistent when so many convenient alibis now exist for painlessly taking away the sins of the world. The extract concludes with a crucial distinction to which Pascal returns many times, as 'external and physical' things are contrasted with the 'internal, spiritual impulse of intention.' The intention may, after all, be so immaculately resplendent that even to an honest man it may seem to justify the rough-hewn methods of its fulfilment. 'The Society of Jesus', writes Edwyn Bevan,[11] 'intent to dominate men for their own good, has been the Church's most effective agent.' In the salvation of a soul, what method is taboo?

and from the salvation of a soul it is but a short step to the defence of one's honour.

This distinction between matter and spirit, violence and truth, *might* and *right*, reaches the height of impassioned eloquence in the twelfth *Provincial Letter*. Here Pascal begins by discussing the somewhat involved question of the attitude of Lessius, a Jesuit professor at Louvain, towards bankruptcy. The whole passage may be cited as an example of the range of Pascal's eloquence and invective, from the close infighting of a particular dispute about one scholar's quotation of another to the majestic utterance of a thinker who views all things from the standpoint of eternity. Just as remarkable as the range of Pascal's eloquence is the rapidity of his transition from small issues to great. But to him, fundamentally, there are no small issues, for – as his opponents, the Jesuits, know only too well – through the apparently small things of the world great ends may be achieved. 'I', he writes (804–5; *190–2*),

> shall waste no time in showing you that Lessius . . . takes undue advantage of the law which allows bankrupts merely a bare livelihood but not a decent standard of living: it is enough that I have justified Escobar against such an accusation. That is more than I had to do. But you, Fathers, are not doing what you ought to be doing: for it is up to you to reply to the passage from Escobar, whose decisions are convenient in that, being independent of what precedes and follows, and in so far as they entirely consist of short articles, they are not subject to your distinctions. I quoted the whole of his passage, which allows *people who make a composition with their creditors to retain enough of their admittedly ill-gotten gains to provide their families with a decent standard of living*. At which I exclaimed in my Letters [745; *123*]: *Fathers, how can that be? By what strange charity would you rather that these assets should belong to those who have wrongfully acquired them than to their lawful creditors?*
> That is the question which must be answered: but it puts you in an awkward position, from which you make pointless attempts to escape by turning the question on its head and quoting other passages from Lessius, ones which are totally irrelevant. So I ask you whether this maxim of Escobar can be followed in all conscience by those who go bankrupt. And mind how you reply. For if you answer no, what will become of your doctor and your doctrine of probability? And if you say yes, I shall report you to the High Court.

> I leave you, Fathers, in this awkward dilemma, for I have no more space here to deal with the next imposture concerning Lessius's passage on homicide; that will be for next time, and the rest later on.
>
> Meanwhile I shall say nothing about the Notices, full of scandalous falsehoods, with which you conclude each imposture: I shall reply to all that in the Letter [XIII] in which I hope to reveal the source of your slanders [805–18; *193–206*]. I pity you, Fathers, for resorting to such remedies. Your insults will not resolve our differences; your various threats will not prevent me from defending myself. You believe that you have might and impunity on your side, but I believe that I have truth and innocence on mine. It is a long and strange war when violence tries to suppress truth. All the efforts resorted to by violence cannot undermine truth: they merely serve to reinforce it. All the enlightenment which truth can bring can do nothing to halt violence and only exasperates it all the more. When might combats might, the greater destroys the less; when words are pitched against words, those that are true and convincing confound and scatter those that are only vanity and lies; however, violence and truth have no power over each other. But let no one claim that they are equal because of this. For there is this huge difference between them, that violence is limited in its course by God's decree as He applies its effects to the glory of the truth it is attacking, whereas truth exists from eternity and will eventually prevail over its enemies, because it is eternal and mighty as God Himself.

The first point at issue in this lofty invective – the apparently, though not really, trivial one – is whether Lessius actually said that bankrupts could keep back enough money in order to maintain a dignified standard of living, or whether he did not. Jacques Nouet, in his *Third Imposture of the Replies to the Provincial Letters Published by the Secretary of Port-Royal against the Fathers of the Society of Jesus*, had tended to give the impression that Lessius never maintained that opinion. With delightful self-confidence Pascal points out, however, that in referring in Letter VIII (745; *123*) to Lessius's remark on bankrupts he had merely quoted the arch-priest of casuistry, Escobar, who himself purported to be quoting from Lessius's *Concerning Justice, Law and the Other Cardinal Virtues*. Did he, Pascal, have to verify Escobar's own quotations? Thus he imprisons not only Lessius and Escobar but all Jesuits in an inextricable dilemma.

For either Lessius did say this about bankrupts, or he did not. If he said it, then the Jesuits are guilty of imposture for denying that he did. If he did not say this, then Escobar is guilty of imposture for quoting Lessius as having said so. In point of fact, Lessius did argue that bankrupts should be allowed to withhold enough from their creditors for themselves and their families to maintain a dignified standard of living: a point Pascal has lost no time in scoring against the Jesuits, even before this extract begins.

Hence the peremptory manner in which Pascal can dismiss the Jesuits and all their tricks in the brief middle paragraph. With matchless self-assurance he has confronted them with a dilemma from which not even they can extricate themselves. Yet, twisting the rapier within the wound, he does not fail to give them a foretaste of the next 'imposture' he will accuse them of. And that will be merely one of a whole series that can be laid to their account: *that will be for next time, and the rest later on*.

The temporal vista, with its obscure and infinitely receding backcloth of deceit and intrigue, is rapidly followed by the timeless one – the paragraph, fittingly, with which Letter XII ends. This is the vista of the world viewed *sub specie æternitatis*, the cosmic onslaught of the 'principalities [and] powers'[12] on the divine Truth: a Truth which is presented as being so wonderful that it is almost co-equal with God Himself, eternal and all-powerful like its Creator. Again, Pascal seems to have been the first writer to have conceived of levelling such a superb accusation against his opponents (though the echoes of Platonic thought are obvious). This counter-attack on the Jesuits is, in fact, made on two levels; and Pascal very skilfully blends and intermingles the two. The first of these is empirical, as objective as any scientific statement; the second, a figure of rhetoric. In the first place, he makes the obvious point that the battle which he and his opponents are fighting is an unreal one, this side of the grave, since neither side can hope to convince or confound the other. Their standpoints are alien; between them there is no common ground; philosophically speaking, it is the juxtaposition of orders which are discontinuous. When brute force encounters brute force, there is at least a recognizable outcome: might is right in such circumstances. When thought clashes with thought, truth will come out in the forum of argument and debate; *magna est veritas et prævalebit*: right is mighty then. But when brute force and thought clash, there can be no convincing victory – might belonging to the temporal world, and *right* to the eternal. Arrayed against him and his friends the Jansenists is all the panoply of political power and clandestine intrigue. He accuses the Jesuits elsewhere of what nowadays would be tantamount to brain-washing, or frighteningly close to the strategy of the big lie endlessly repeated: 'It is time for me to put an end once and for all to your audacity in calling me a heretic, an audacity that grows day by day', he writes in Letter XVII (866; 259) to François Annat, the foremost of all French Jesuits and confessor to Louis XIV.

> You do this to such an extent in this book which you have just published[13] that it has become intolerable, and I should eventually incur suspicion if I did not answer as a charge of this kind deserves. I had despised such an insult when it occurred in your colleagues' writings . . . My Fifteenth Letter was a sufficient reply; but you now speak about it differently, making it in all earnestness the crux of your defence; it is virtually the only argument you use. For you say *that to answer my fifteen letters it is only necessary to say fifteen times over that I am a heretic; and that, once having been said to be such, I do not deserve to be believed by anyone.* In a word, you treat my apostasy as if it were beyond question: you take it to be a firm premise upon which you boldly build.

As a scientist and a logician, Pascal objects with all the vehemence at his command to any method of discussion where there is no meaningful meeting of minds. In terms of propaganda and psychological warfare, he and his friends are in any case hopelessly outmatched by the Jesuits. His *Letters* have to be issued clandestinely; Annat's book is published with the King's approval! But it does not follow from this that Annat's book is right, and his own views wrong. Towards the end of the *Letters* it is clear that Jansenism cannot (in the short run, at any rate) hope for any worldly triumph in its clash with the Jesuits. By the end of the controversy, as the last part of the extract from Letter XII shows, Pascal echoes the attitude of Jesus towards Pilate: *My kingdom is not of this world*[14] – for the world may despise the truth, but nevertheless the truth is eternal and God-given. This is the second aspect of the self-vindication to François Annat, the figure of rhetoric stemming no doubt from his own unshakable belief in the rightness of his opinions, but certainly not demonstrable by any objective criteria. Granted that might and thought can never meaningfully collide, it does not automatically follow that thought is right. Yet Pascal boldly arrogates to himself and his Jansenist friends the certainty that, however much they may be crushed and downtrodden in the arena of religious politics, they and they alone represent the Truth which is eternal and divine. Nor, in this meeting of disparates, is it simply a case of honours even. The Jesuits' might may triumph in one sphere, the Jansenists' right in another. But right is as overwhelmingly superior to might as the eternal is to the temporal. What is more, God will only countenance might's triumph for a little day; and ultimately will turn even that short-run triumph of temporal violence to the greater glory of the eternal Truth. As an arraignment of the Jesuit standpoint, Pascal's invective is notable both for its unself-questioning self-assurance and for the dignified and restrained nobility of its language. The case he presents for Jansenism in the Twelfth and Seventeenth Letters would not of itself stand up to a rigorous examination by either scientists or lawyers. Yet as a figure of majestic rhetoric, Pascal's apologia remains: a matter not of the mind alone but of the heart, not of logic but of charity; the embodiment of a poetic truth higher than any truth of law or physics. No one before Pascal's time had ever assumed the mantle of eternal Truth with so much eloquence and passion.

On the level of the Church militant, for the greater glory of the Church within the world, he accuses the Jesuits of various devious opportunist tactics – the end justifying the means. But on the level of the divine Truth, which he claims to represent, can he likewise be accused of various devious artifices of argument? Can he be called unscrupulous at times in the weapons he employs against his enemies, albeit for the sake of Christ risen, ascended and glorified? Foremost amongst the charges of unscrupulousness levelled against Pascal is that, also in Letter XVII, he carefully dissociates himself from Port-Royal when, defending himself against Annat's accusation of heresy, he demands the tangible proof. 'When have I been seen at Charenton?'[15] he asks.

> When have I been absent from mass, when have I failed in my Christian duty towards my parish? When have I done anything to act in concert with heretics, when have I been in schism with the Church? What Council have I contradicted? What Papal constitution have I violated? You must reply, Father, or . . . you know exactly what I mean. And what is your reply? I ask everyone to take note. First of all you assume *that the writer of the Letters is from Port-Royal*. You then say *that Port-Royal has been*

> *declared heretical*; from which you conclude *that the writer of the Letters has been declared heretical*. So it is not on me, Father, that the onus of this charge falls but on Port-Royal; and you lay it against me only because you assume that I am one of them. Thus I shall not find it very hard to defend myself, as I have only to tell you that I am not one of them and refer you to my Letters, in which I have said *that I am alone* and, quite explicitly, *that I am not from Port-Royal*, as I did in the Sixteenth Letter [848–9; *239*] which came out before your book (867; *259–60*).

Yet it is strictly true to say that Pascal did not belong to Port-Royal. Indeed, as has been noted, he may only ever have paid two visits to Port-Royal des Champs, whilst a total of five visits during his lifetime – when sometimes he may have stayed at Vaumurier – would seem to be the absolute maximum. Although he sympathized with the Jansenists as devout holy men and his personal friends, he may never have fully subscribed to Jansenism in the theological sense. It was for this reason that he could preserve his anonymity as the writer of the *Letters*, an anonymity not officially broken until after his death. The opening of Letter VIII pokes fun at the difficulty his opponents were having in piercing the mystery of his identity:

> Some think I am a doctor of the Sorbonne: others ascribe my letters to four or five people who, like me, are neither priests nor churchmen. All these false suspicions bring home to me the fact that I have been quite successful in my plan of being known only to you, and to the good Father who still puts up with my visits and whose conversation I still put up with, albeit with great difficulty (740; *117*).

The success of his incognito stemmed from the fact that he was seldom, if ever, seen at Port-Royal des Champs, not known as a writer, and not closely associated with Jansenism in the public mind. His visits to Port-Royal de Paris were, of course, much more frequent; but that, the Jesuits must have assumed, was more to see his sister than to consort with heretics.

In his quotations from his opponents Pascal is also remarkably fair.[16] The more far-fetched and unlikely of these, if traced back to their sources, are on the whole either perfectly or near-perfectly literal: even, for instance, when the narrator and the Jesuit priest are discussing the bribery of judges. 'Can you not understand', asks the Father (749; *128*),

> that a judge is bound to render justice, and thus he cannot sell it; but he is not bound to render injustice, and thus he can take money for that? So it is that all our major authors, such as Molina, disputations 94 and 99; Reginaldus, book X, numbers 178, 184 and 185; Filiutius, treatise XXXI, numbers 220 and 228; Escobar, treatise III, exercise i, numbers 21 and 23; and Lessius, book II, chapter xiv, disputation 8, number 52, all consistently teach *that a judge is certainly bound to repay what he has accepted in order to render justice, unless it was given to him out of generosity: but that he is never bound to repay what he has accepted from someone in whose favour he has pronounced an unjust judgment.*

This astounding topsy-turvydom is a literal quotation from Lessius's *Concerning Justice, Law and the Other Cardinal Virtues*, except that the exact reference should read: 'book II, chapter xiv, disputation 8, number 55.' Likewise in the discussion, in Letter VI (716–17; *89–90*), of the occasions when members of a religious order are permitted to leave off their monastic habit:

> 'Popes have excommunicated monks who leave off their habit, yet our twenty-four Seniors[17] nevertheless speak thus, in treatise VI, examination 7, number 103: *On what occasions may a monk leave off his habit without incurring excommunication?*[18] Several [favourable circumstances] are quoted, including the following one: *If he leaves it off for a shameful reason, like going out to steal, or going incognito to places of sin, intending to put it on again soon afterwards.*[18] So, quite obviously, the Bulls ignore such cases.'
>
> Finding this hard to believe, I asked the Father to show it to me in the original; and I saw that the chapter in which these words occur is entitled 'Practice according to the School of the Society of Jesus'; 'Praxis ex Societatis Jesu Schola';[19] and I read these words: Si *habitum dimittat ut furetur occulte, vel fornicetur.*[20] And he pointed the same thing out to me in Diana, in these terms: Ut eat incognitus ad lupanar.[21]

The first two of these quotations are meticulously copied from Escobar, who, without being specific, cites the Spanish Jesuit Tomas Sánchez as his authority. His last, from Antonino Diana, is again word-perfect except for the addition of the one word *incognitus*:[22] this merely reinforces, without extending, Diana's meaning since all monks taking off their monastic habits before visits to brothels are perforce going there unrecognized.

Similarly, Pascal is reasonably fair – albeit a trifle inaccurate now and then – in his bizarre concatenation of the names (all foreign-sounding) of the world's new religious authorities. These are the aggressively modern innovators who are constantly refurbishing and refining the old lack-lustre doctrines of the Evangelists and Fathers of the Church. 'In other words, Father', says the narrator (713–14; *86*),

> 'your arrival has meant the disappearance of St Augustine, St Chrysostom, St Ambrose, St Jerome and the others so far as morality is concerned. But do at least let me know the names of the men who have succeeded them: who are these new authors?'
> 'They are very able and famous men', he replied. 'They are Villalobos, Coninck, Llamas, Achokier, Dealkozer, Dellacruz, Veracruz, Ugolin, Tambourin, Fernandez, Martinez, Suarez, Henriquez, Vasquez, Lopez, Gomez, Sanchez, de Vecchis, de Grassis, de Grassalis, de Pitigianis, de Graphæis, Squilanti, Bizozeri, Barcola, de Bobadilla, Simancha, Perez de Lara, Aldretta, Lorca, de Scarcia, Quaranta, Scophra, Pedrezza, Cabrezza, Bisbe, Dias, de Clavasio, Villagut, Adam a Manden, Iribarne, Binsfeld, Volfangi a Vorberg, Vosthery and Strevesdorf.'[23]
> 'Oh Father!' I said, feeling quite alarmed, 'were all those men Christians?'
> 'What do you mean, Christians?' he retorted. 'Was I not telling you that they are the only people through whom we govern Christendom today?'

The two historical allusions which provoked the fiercest outrage from Pascal's opponents are also strictly accurate. The first of these, alluding to the so-called *Mohatra contract* (743–4; *121–2*), refers to a device thought up in the sixteenth century for circumventing the Church's age-old prohibition of usury; perhaps because of Pascal's condemnation of it, it was to be condemned by the Holy See within twenty years of his death.[24] In Letter VIII (744; *121*) he quotes Escobar as defining it thus: 'The Mohatra contract is one whereby materials are bought for a higher sum and on credit, only to be sold simultaneously to the same person for a lower sum and for cash':[25] except that it omits the reference to 'a lower sum', the quotation is accurate. When asked by the narrator whether any other casuist has referred to the Mohatra contract, the Jesuit priest points to the recently published *Digest or Complete Compendium of Summas*, by the Spanish Franciscan friar Juan Soria-Butron, reciting its definition of the bargain:[26] the quotation is accurate once again. When, however, Pascal goes on to make the really substantial point that these casuists are giving their blessing to the Mohatra contract even though it is illegal, he departs slightly but noticeably from his usually high standards of verbal literalness. The gist of his argument, that under the Jesuits' leadership the moral guidance of the Church is encouraging men to defy the laws of the State, is truthful; but he is rather free in his rendering of a passage taken from Lessius's *Concerning Justice, Law and the Other Cardinal Virtues*[27] – not perhaps distorting the spirit of Lessius's moral instruction but certainly destroying the letter: paraphrasing in order to point up the fundamental laxity (744; *122*).

The notorious Jean d'Alba controversy, sparked off by Letter VI, is the second of the two historical allusions which goaded the Jesuits to fury. In his most characteristic and deceptively naive way, Pascal leads up to the story of the servant prosecuted for stealing from his Jesuit employers by means of a quotation (724–5; *99*) from Étienne Bauny's *Compendium of Sins*: 'If menservants complain about their wages, may they increase them themselves by laying their hands on as much of their masters' property as they think is necessary in order to make the said wages equal to their toil? They may do so in certain circumstances, as when they are so poor, when looking for a job, that they have had to accept whatever was on offer, whereas other menservants at their level of employment earn more elsewhere.' This quotation is a practically literal reproduction of the original, except that in the *Provincial Letters* the reply to Bauny's rhetorical question summarizes a considerably longer statement.[28] The narrator then goes straight to the heart of the matter with an account of d'Alba's fate (725–6; *99–100*):

> This unfortunate man confessed under interrogation that he had taken a few pewter dishes from your Fathers; . . . quoting as his justification this doctrine of Fr Bauny, which he presented to the judges together with a document from one of your Fathers, under

> whom he had studied cases of conscience and who had taught him the same thing. Whereupon M. de Montrouge[29] . . . *expressed the opinion that on the basis of some of the writings of these Fathers, containing an unlawful and pernicious doctrine contrary to all natural, human and divine laws, and capable both of causing the downfall of any family and also of justifying any domestic theft, he did not consider that the accused should be acquitted. But rather that, in his opinion, this unduly loyal disciple should be whipped outside the College door by the public executioner, who should at the same time burn the writings of these Fathers dealing with larceny, and that they should be forbidden ever again to teach such doctrines on pain of death.*

The man telling the narrator and others this story

> added that M. de Montrouge's judgment is recorded in the official records of the Châtelet[30] for all to see. We enjoyed this tale.'
> 'What's so funny about that?' asked the Father. 'What does it all mean? Here I am, telling you all about the maxims of our casuists; I was just about to deal with the maxims concerning the nobility, but you keep butting in with stories that are beside the point.'
> 'I only mentioned it by the way', I said, 'and also to remind you of an important aspect of the subject, which I think you overlooked when working out your doctrine of probability.'

The aspect of the matter overlooked by the Jesuits was, of course, that doctrines which they could establish as probable – concerning, for instance, the bribing of judges or the stealing of one's master's possessions or even the committing of murder – were still contrary to the law of the land. Probabilism could take away the sins of the world, but not the *crimes*. 'I told you, Father, that you would never get anywhere unless you had the judges on your side', the narrator quips in the Seventh Letter (736; *112*). And again, in Letter VI (726; *101*):

> if on the one hand you are the confessors' judges, are you not also, on the other hand, the judges' confessors? You wield very great power: compel them, on pain of exclusion from the sacraments, to acquit criminals who have a probable opinion in their favour, so that it does not happen that people whom you exculpate in theory are either whipped or hanged in practice – something which would bring probability into great scandal and contempt.

Thus Pascal implies, through this remark of the narrator to the priest, that the Jesuits – potentially, at least – are threatening to subvert the whole fabric of the State. Yet the law remained obdurately resistant to their corrosive influence. The Jesuits might well encourage crime by stealth on the ground that it was not sin; but crime it still was, to be punished as such by the judges. Pascal's vast superiority over his opponents in this respect was that he could, and did, truthfully quote chapter and verse in support of his indictment of them. There was no getting away from the fact that the Jesuits' writings had condoned homicide, larceny, bribery and embezzlement. In his sadly unequal battle against unscrupulous enemies, Pascal was honest. The most that can be said against him is that, although his quotations from the Jesuits are on the whole painstakingly precise, he was always pointing to extremes, the furthest lengths to which their doctrines could logically be extended.

Needless to say, the Jesuits' leaders retaliated savagely to the mounting storm. The Jean d'Alba story was substantially true, Claude de Montrouge's judgment accurately reported, the trial date correctly indicated, yet nevertheless Pascal's account contained certain slight but artistically lethal distortions. It seems that not only did Jean d'Alba nurse a grievance about his wages; he could also apparently prove that his employers owed him ninety francs. And, conversely, he was suspected not merely of a 'domestic theft' but also of stealing chalices from the College chapel. Thus, his crime was greater than Pascal will admit, whilst his trial did not turn merely on Étienne Bauny's justification of servants who steal in order to make up their wages to an acceptable level. Nor was Jean d'Alba's punishment as sensational and as damning for the Jesuits as the *Provincial Letters* suggest. Rather than being whipped outside the main entrance to the College, the offender, having duly been found guilty by the court, was simply ordered to return home to the village of his birth. Letter VI, whilst conveying the essential truth about the d'Alba trial, tricks it out nevertheless in a more vivid and dramatic form.

Wherever Pascal could not be directly controverted – his facts and quotations disproved, his honour impugned – it was, of course, always possible to accuse him of disrespect for the Church and sacred things; and this, in fact, was one of the Jesuits' most frequent lines of attack. 'If you spoke like that in places where you were not known', the priest cautions the narrator, 'there might be people who would take your remarks amiss and accuse you of ridiculing religious things' (750; *129*). Ridicule was Pascal's deadliest weapon, though it is still untrue for Gide to claim that nobody would read the *Provincial Letters* any more *if* they were serious. For behind all the fantasy and wit there lies a deadly seriousness, the seriousness that can be engendered only when issues of life and death are at stake. Nowhere, perhaps, is this unique distillation of reportage and fancifulness more cunningly blended than in the passage discussing whether priests and monks may commit murder (737–8; *114–15*), which is whimsical satire in all earnestness.

> 'Indeed, according to our celebrated Fr L'Amy,[31] priests and monks are even allowed to strike first against people who are wanting to besmirch and defame them, by killing them so as to prevent that.[32] But only if the intention is properly directed. Here are his words, volume V, disputation 36, number 118: *'It is permissible for a priest or monk to kill any slanderer who threatens to publicize scandalous crimes concerning either his Community or himself if that is the only means of preventing him, granted that he is about to give circulation to his calumnies unless he is promptly killed. For, in this case, just as it would be permissible for such a monk to kill anyone wishing to deprive him of his life, so likewise it is permissible for him to kill anyone wishing to deprive either him or his Community of their honourable name, in the same way as in the secular world.'*[33]
> 'That's news to me', I replied. 'I believed the exact opposite, without giving the matter any thought, because I had heard that the Church feels such abhorrence for bloodshed that she does not even permit ecclesiastical judges to be present at criminal trials.'
> 'Don't be put off by that', he said. 'Our Fr L'Amy is very good at proving this doctrine although, in a gesture of humility entirely befitting this great man, he submits it to his readers' discretion. And Caramuel, our illustrious defender, quoting it in his *Fundamental Theology*, page 543,[34] believes it to be so certain that he argues *that the contrary is not probable*; and from this he draws admirable conclusions, such as this, which he calls *the conclusion to end all conclusions, 'conclusionum conclusio': That not only may a priest kill a slanderer under certain circumstances, but even that there are certain circumstances where there is a duty upon him to do so: etiam aliquando debet occidere.*[35] He examines several new questions on the basis of this principle, including the following one, for example: MAY THE JESUITS KILL THE JANSENISTS?'
> 'Oh Father!' I exclaimed, 'that is a very surprising point of theology! I reckon the Jansenists are already as good as dead on

> the basis of Fr L'Amy's doctrine.'
> 'That is just where you are wrong', replied the Father. 'Caramuel concludes the opposite from the same principles.'
> 'Well, Father, how does he manage that?'
> 'Because', said he, 'they do our reputation no harm. Here are his words, numbers 1146 and 1147, pages 547 and 548: *The Jansenists call the Jesuits Pelagians; can one kill them for that? No, inasmuch as the Jansenists no more dim the splendour of the Society than an owl dims the splendour of the sun.*'

Against such brilliantly lavish scorn there was indeed no other possible line of defence for the Jesuits than to say that Pascal 'ridiculed sacred things' (779; *163*).

What in reality Pascal was doing, though his opponents could not or would not realize it, was to ridicule their travesty of the Church by reference to things that were truly sacred. Sometimes he did this by direct quotation from their writings, sometimes by a flight of imaginative fancy: but always by juxtaposition. 'I can quite see', the narrator admits (714; *87*),

> that anything is acceptable to you except for the ancient Fathers, and that you are masters of the field. You have only to keep pressing on.
> 'But I foresee three or four great inconveniences, and powerful obstacles standing in your way.'
> 'What?' asked the Father, quite astonished.
> 'There are', I answered, 'Holy Scripture, the Popes and the Councils, which you cannot disown, and which all follow the path of the Gospel.'
> 'Oh, is that all?' came his reply. 'You did give me a fright . . . '

And (737; *113*): 'Look all through the ancient Fathers to see for what sum of money it is lawful to kill a man. What will they tell you but: *Non occides*; Thou shalt not kill.' 'The Fathers were all right by the moral standards of their times; but they are too remote from ours' (713; *85*). *'It is permissible to kill someone for the value of a crown, according to Molina'*[36] (737; *114*). In countries where the mystery of the Incarnate God crucified, dead and buried would not be reverently accepted, missionary Jesuits 'suppress the scandal of the Cross' (705; *76*). As for praying for *the instant death of people who are preparing to persecute us*, an accurate quotation from Pedro

Hurtado de Mendoza,[37] the Church has not (yet!) included in the prayer book 'everything that can be asked of God. Besides, that was not possible; for that opinion is newer than the breviary: you are no good at chronology' (730; *105*). Both the rich man's desire to give alms and his desire to avoid doing so 'are safe according to the same Gospel; one, according to the Gospel in the most literal and straightforward sense;[38] the other, according to that same Gospel as interpreted by Vasquez'[39] (716; *89*). 'When I tell you that our Fathers have reconciled these things [the Gospel law and the world's laws], all you can say is that you are astounded' (728; *103*).

Such fanciful play with a soberly documented historical and theological situation is sustained throughout by an irresistible verve of wit, farce, punning, paradox, antithesis and hyperbole: the joke about Aristotle (701; *71*), the bravura passage (692; *60*) on Arnauld whose single but insuperable fault is to be himself, the farcical remission from fasting (707–9; *78–80*), the pun (675; *40*) about proximate powers and one's neighbour (*prochain* in both cases), the parable (680–1; *46–7*) of the doctors advising the traveller attacked by robbers, the hyperbolical tirade (683; *49–50*) on the Dominicans' failure to defend Efficacious Grace, the cool irony of 'people seldom think of murdering anyone except their enemies' (732; *108*), the neat antithesis (689; *57*) that 'it is much easier for them to find monks than arguments', and the paradoxical 'I am not even sure whether a man would not feel less resentment at being brutally killed by hotheads than at feeling he was being stabbed to death by devout people for conscientious reasons' (739; *116*). There is the ambling mumbo-jumbo of *'probabiliter obligatus, et probabiliter deobligatus'* (723; *97*), and the darting ferocity of 'I believed you could only take away sins; I did not think you could introduce them too' (713; *85*).

It is this stark contrast between the sobriety and rapier-like incisiveness of Pascal's diction and the lush turgescence of his opponents' which explains his method: the contrast between Classicism and the Baroque. Their inflated periods conceal an intellectual void, which his terse precision detects. He has only to let them talk on, in their own (more or less) faithfully transcribed words, for them to talk themselves out of favour.

> *He has no eyes for the beauties of art and nature. He would believe he had taken on an awkward burden if he found pleasure in anything. On feast days he withdraws to the graveyard. He would rather be in a tree-trunk*

> *or a cave than in a palace or on a throne. As for insults and injuries, he is as insensitive to these as if he had the eyes and ears of a statue. Honour and glory are things unknown to him, idols to which he has no incense to offer up. To him a beautiful woman is as a phantom. And those haughty and regal faces, those charming tyrants who everywhere make willing conquests, slaves without chains, have the same power over his eyes as the sun has over an owl's*:

these words from the Jesuit Fr Pierre Le Moyne's *Moral Portraits* (757; *136–7*)[40] describe – and denigrate – the 'savage' (756; *136*), or uncouth man 'incapable of feeling natural and decent affections';[41] the savage displays in his life an excess, indeed a perversion, of the virtue of temperance. This cento of quotations is the acme of religious worldliness. For, implicitly at least, Le Moyne appears to be denigrating the austere religious life by putting it on a level with that of the savage. Religious self-abnegation is, it seems, more the result of a person's natural disposition or temperament than of any strenuous effort to practise piety. With reproachful impatience Le Moyne condemns the savage for offering up no incense to the idols of honour and glory. Putting words into his own Jesuit father's mouth, Pascal ironically dismisses the savage's (or the saint's?) way of life as 'the ridiculous and brutish ways of a melancholic madman' (756; *136*). To Le Moyne's bombast and insipid preciosity his terse astringency is an effective rejoinder: 'if this is the picture of a man totally detached from the feelings which the Gospel bids us renounce, I confess I can make no sense of it' (757; *137*).

The preciosity and flowery rhetoric of Pierre Le Moyne do not exclude prurience. 'It may be permissible to dress up at an age which is the flower and prime of life', he writes in his *Easy Piety*,[42] in a cento (763; *144*) also gathered together by Pascal in Letter IX. 'But that must be as far as we go: it would be strangely inappropriate to look for roses in the snow. It is only for the stars to be always dancing,[43] because they have the gift of eternal youth. The best thing, as far as this is concerned, would therefore be to consult reason and a good mirror, to yield to propriety and necessity, and to withdraw as night approaches.' This is a worldliness which, even when thinking of the stars, cannot resist the metaphor of the ballroom. Not even the Virgin Mary could escape from the cloying fulsomeness and unwholesome attentions of Jesuit writers. Paul de Barry, in his *Paradise Opened to Philagie by Means of a Hundred Easily Performed Devotions to the Mother of God,* writes of her in terms more appropriate to Mary Magdalene than to a 'Virgin undefiled.' All hundred devotions are easy, the Jesuit father explains in another of Pascal's centos (753–4; *132–3*), this time a confection from Barry:[44]

> '*Salute the Holy Virgin when you come across images of her; recite the little rosary of the ten pleasures of the Virgin; frequently utter the name of Mary; charge the Angels with paying her our respects; desire to build more churches to her name than all the monarchs in the world have done; wish her 'good day' every morning and 'good evening' as night draws on; and say the Ave Maria every day in honour of the heart of Mary.* With such devotions, he says, you can be sure of winning the Virgin's heart.'
>
> 'But, Father', I interjected, 'only if you give her yours too?'
>
> 'That is not necessary', he replied, 'when you are too closely attached to the world.'

Mary is presented by Barry as an earthly virgin to be wooed and, worse still, deceived by trinkets, baubles and sham displays of affection: naive, credulous, an *earthly* virgin easily conquerable by a *worldly* man. '*And now say that I do not supply you with easy devotions for winning the favours of Mary*', Pascal quotes Barry as saying (again, with perfect accuracy), to which he replies, with sexual innuendo: 'That is facility itself' (754; *133*). The seven words 'only if you give her yours too' epitomize his rooted objections to the Jesuits: that, far from himself ridiculing sacred things, they are degrading them – defiling even the Virgin by glossing her love for mankind with sexual overtones and, above all, making her love mechanical: a sort of magic superstitious idolatry conveniently dispensing salvation at the turn of a well-worn formula. Exactly the same objection applies to their view of the sacraments (849–52; *239–43*), their doctrines of mitigation of confession (765–6; *147*) and automatic absolution (770–1; *153*), and their trivialization of the Eucharist (764; *145*):

> 'But there is another useful thing in our learned Turrianus, *Selections*, part II, disputation xvi, doubt 7: *That one can hear half of one priest's mass, and then the other half of another's, and even that one can first of all hear the end of one and after that the beginning of another.*[45] And I will tell you something else as well: it has also been declared permissible *to hear two halves of mass at the same time*

said by two different priests, one of them beginning the mass when the other has reached the Elevation; because it is possible to pay attention to both sides at the same time, and two halves of a mass make up one whole: Duæ medietates unam missam constituunt.[46] This is what has been decided by our Fathers Bauny, treatise VI, question 9, page 312[47]; Hurtado, *On Sacraments*, volume II, *On the Mass*, disputation v, difficulty 4; Azorius, part I, book VII, chapter iii, question 3;[48] Escobar, treatise I, examination ii, number 73, in the chapter on 'The Practice of Hearing Mass According to our Society.' And you will see, in this same book, in the editions published at Lyons in 1644 and 1646, what consequences he derives from all this when he writes as follows: *I conclude from this that you can hear mass in a very short time: if, for example, you come across four masses being said simultaneously, and so arranged that just as one is beginning, another has reached the Gospel stage, whilst a third has got as far as the Consecration and the fourth has reached the point of Communion.*'[49]
'To be sure, Father, mass at Notre-Dame will be over in a minute by this method.'

In Pascal's view, mechanical observance is not enough; what counts, and what is to be hoped for and encouraged in man, is a sincere heartfelt turning to God, a love of Him Who 'so loved the world, that He gave His only begotten Son.' Quoting these words from John III 16 in his final scornful dismissal of the Jesuit priest at the end of the Tenth Letter, he underlines the cheap shallowness of their view that 'the world, redeemed by Him, shall be exempted from loving Him!' (778; *162*) Even more than their teaching on probability and sufficient grace, their doctrine of attrition merits his deepest contempt. He cannot accept that a man need only be motivated by the fear of hell-fire in order to win his salvation, rather than come humbly to God out of a contrite heart, longing for the salvation that will eternally unite his love for his Maker with God's love for him. When 'the human mind makes such insolent sport of the love of God' (776; *159*), it can stoop no further. 'Our Fathers', inanely boasts the Jesuit priest (776–7; *160*), 'have released men from the *tiresome* obligation of actually loving God.[50] And there are so many advantages to this doctrine that our Fathers Annat, Pinthereau, Le Moyne and A. Sirmond himself have vigorously defended it against attempts to attack it.' 'Thus', and the last word is Pascal's (778; *162*), 'people who have never loved God in all their lives are made worthy by you of enjoying God in all eternity.'

From a strictly political and opportunist standpoint the Jesuits were, of course, substantially on the right tack. To lighten God's yoke is the surest means of making, and keeping, the largest number of nominal Christian worshippers. And this Pascal realized, when he made his Jesuit priest say: 'Men are so corrupt nowadays that, since we cannot make them come to us, we really have to go to them' (720; *93*). But he, unlike them, failed to detect any value in a mere religion of formalism and social propriety, the shadow and not the substance of religious worship, devoid of any sense of the supernatural. 'As their morality is wholly pagan', says the narrator's Jansenist friend in Letter V (707; *77*), 'natural powers suffice for its observance.' Hence Pascal's belief in the need for divine support of the fallible human will, hence too the acrimonious controversy about grace. Nevertheless, the doctrine of efficacious grace and the Elect was not too likely to commend itself to that vast majority of human beings whose will, by definition, was weak, whose judgment was fallible, and whose unredeemed nature was corrupt. Pascal was defending a lost cause. Not even his brilliance and tenacity could turn the tables on opponents who were so completely the masters of the tactical opportunity. Intellectually the victory was his; politically he had encountered failure. Arnauld, condemned both in *fact* and in *right*, had for the second, and not the last, time in his life gone into deepest hiding; there was no need for a Nineteenth Letter.

Pascal's personal (though anonymous) condemnation was quick to follow. On 6 September 1657, five months after the publication of the Eighteenth, the *Provincial Letters* were placed on the *Index Librorum Prohibitorum* by Pope Alexander VII. For Pascal, as an Ultramontanist at heart rather than a Gallican, a man believing in the supremacy of the Pope over the Church rather than in the semi-autonomy of the Church in France, this was a bitter blow. In the first of his two letters to Annat, after declaring that he did not belong to Port-Royal, he had made a clear affirmation of religious loyalty: 'my only allegiance on earth is to the Roman Catholic and Apostolic Church, within which I wish to live and die in communion with the Pope, its sovereign head, and outside which I am fully convinced that there is no salvation' (867; *260–1*). Although he had not been excommunicated, his first writings on a religious subject had been declared heretical. For all the work's inherent brilliance and public acclaim, such was the bitter-sweet taste of failure. Even today Molinism, the doctrine of sufficient grace, and Probabilism are the approved doctrines of the Roman Catholic Church.

In an eternal perspective, however, Pascal believed that his was the just – and would be the winning – cause (1073). He is even the winner of the temporal argument in the longer run, at any rate so far as intellectual consistency and moral credibility are concerned, though perhaps a reading of the *Provincial Letters* in 1995 will secure few converts. And all this is chiefly due to his skill and brilliance as a polemist, the inward strength of conviction inspiring the outward eloquence. As an essay in polemical satire, the *Provincial Letters* were a unique phenomenon. Never before had a satirist had so specific a target in view, ridiculing it by so dexterous a combination of realism and fantasy. The reading of the *Provincial Letters*, if not an uplifting religious experience, is certainly a reminder of the versatility of human genius.

The *Letters* are an unfinished and episodic achievement, impossible to discuss from the conventional critical standpoint of a generalized structure. The irony, indeed, of Pascal's literary career is that his two major works, both masterpieces of unusual distinction, are both of them unfinished and in a sense fragmentary. But if the *Provincial Letters* are a reminder of human genius, it is the combined genius of many minds: the genius for folly of Escobar, Molina, Le Moyne, Barry, Diana, Filliucci and numerous other casuists; the scatological genius of Arnauld and Nicole collecting their scabrous specimens in many dark corners, but notably in Escobar; and the arranging and conflating genius of Pascal himself, coupled with his irony, his gift for the terse phrase, the judicious quip, the incisive stab, and the decisive rejoinder.

Sometimes the casuists speak for themselves, always lamentably then. Sometimes – more often indeed – a puppet Jesuit speaks with the idiocy of Pascal's own imaginative verve. Sometimes, but only in the very earliest Letters, the narrator's Jansenist friend contributes a piece of wisdom. Monsieur N, a sympathizer with the Jesuits, defends his friends but never very successfully. With a skill in reportage well ahead of his times, Pascal, journalist-like, rushes from place to place trying to establish the reasons for all this much ado about nothing. He sets down his findings in plain unvarnished French for the cool appraisal of his unbiased correspondent. The turgid rhetoric of his opponents is laughable beside his own sobriety of word and judgment. As a method of denigration, this is essentially truthful. Underlying all the farce, facetiousness and levity is a deep and fundamental seriousness, such as any reading of d'Alba's story would induce.

It is the seriousness of those who fight not only for their lives, reputation and integrity but in defence of a passionately upheld belief. And it is as present in the earlier Letters, beneath all the comedy of the bumbling casuist, as it is in the collective appeals to the Society of Jesus and in the final, despairingly indignant letters to François Annat. Using the world's weapons to condemn worldliness, the *Provincial Letters* are infinitely more serious than their mordant wit and flippant irony would suggest. Yet to the worldly Jesuits Pascal's layman's gift for unmasking imposture, making the abstruse interesting and appealing beyond the theological faculty to the widest public made him not serious enough!

Hilaire Belloc, noting that the '*Provincial Letters* have been in the past unceasingly used, and are even still used, as a weapon against the Catholic Church', has been the fiercest critic of their treatment of the Jesuits. In what is, he admits, 'a cold way of meeting such excellent writing',[51] he regrets that 'of the thousands of Casuist decisions arrived at by a vast number of professors, regular and secular, Pascal chooses to speak only of the Jesuit decisions' and from the latter 'selects what are in appearance . . . only 132,[52] and in real numbers – if we exclude repetitions – only 89.'[53] A 'just analysis' of these 89 seems to him to leave fourteen in contention, 'of which eight[54] were, at one time or another, finally condemned at Rome'[55] whilst another three (725–6; *99–100*. 760; *140*.[56] 760; *141*)[57] are said to be 'capable of confusion with condemned propositions.'[58] Those dealing with financial inducements to judges (741–2; *118–19*),[59] the Mohatra contract (previously discussed) and Adam Tanner's[60] and Gregorio de Valencia's instructions on simony (720–1; *94*) are, in Belloc's view, 'doubtful'[61] cases. Concerning the Mohatra contract, even he is ready to acknowledge that 'Pascal was probably right, and Escobar was probably wrong.'[62]

The fact is, however, that many more than eight points of Jesuit casuistry have, since 1657, been 'condemned at Rome.' Abuses to which Pascal drew attention, and which were subsequently condemned, include double payment for masses (721; *95*), the murder of slanderers (806—807; *194*), gluttony (759; *140*), desiring a father's death (730; *106*) and theft (746; *124*): all these casuistical interpretations of moral conduct were to be proscribed, either in 1665 or 1679, in Papal decrees.[63] Furthermore, Pascal is surely to be congratulated – rather than, by Belloc, condemned – for the fact that the eight (in reality, more than eight!) casuistical teachings which later were censured by popes cannot now 'be brought up in accusation against the moral system of the Catholic Church.'[64] In some things, however, he is undoubtedly inaccurate (in quotations from Valencia[65] and Lessius,[66] for instance) but these, within the total situation he is describing, are of small account. Contrary to Belloc, any 'just analysis' of the *Provincial Letters* would suggest that its quotations, translations and digests are in the main exceedingly reliable.[67]

It is also true, in Kenneth Kirk's words, that 'Vasquez, the first systematic probabilist, . . . [insisted] upon the safer course being taken where there is danger of a breach of the natural law or of charity, or in the case of the sacraments' and that 'even the notorious Escobar had limited the use of probable opinions to cases in which no danger threatens which prudence, justice or charity bids us avoid.'[68] This was the theoretical position. But in every department of life, and not least perhaps in moral theology, theory and practice can be far apart. As had been recognised well before the decade in which the *Provincial Letters* were written,[69] Escobar's Twenty-Four Seniors had indeed opened up the way to moral permissiveness.

With 'so much wit and fervour' (in Belloc's words)[70] Pascal stresses what *might* be perpetrated by casuists, Jesuits or otherwise, under the ægis of the doctrine of probabilism; and furthermore that this, unless eradicated, might *increasingly* be perpetrated in the future. At a literary and philosophical level his device of the Jesuit Father compensates for any shortcomings in generalization. Historically speaking, the fact that so many of the matters censured in the *Provincial Letters* were later to be censured in Papal decrees serves to remind us that Pascal's uneasiness concerning the laxities of probabilism was well founded.

6

The Christian Life

The quarrel with the Jesuits' moral slackness in the confessional continued long after the publication of the eighteenth *Provincial Letter* and the work's condemnation by Pope Alexander VII. Where Pascal and his Jansenist friends had begun, the whole Assembly of the French Clergy now took over. Significantly, the parish priests of France voiced a deep objection to the activities of an international religious movement, organized on semi-military lines, which had so much less responsibility than they for the guidance and moral welfare of simple people. As early as May 1656 the parish priests of Paris had urged the Assembly of the Clergy to take steps to condemn the Jesuits' lax probabilism. They were followed in this by the parish priests of the dioceses of Rouen and Amiens. So much for the practical impact of the fifth to the tenth *Provincial Letters,* all published between 20 March and 2 August 1656, in which the narrator and Jesuit priest between them expose the rottenness at the heart of Jesuit morality. At first, however, the bishops and the Assembly of the Clergy merely ordered (in February 1657) the reprinting and circulation to all parish priests of the *Instructions for Confessors* composed by Carlo Borromeo almost a century earlier when he himself had been appalled by the havoc wrought by the Jesuits in his own archdiocese of Milan. Then occurred a new sensation, with the publication, in December 1657, of Georges Pirot's *Defence of the Casuists*. By far the most outrageous act of the Jesuits in France, it was a book which expressed no regret for the scandalous teachings of Bauny, Caramuel, Diana, Lessius, etc. denounced in the *Provincial Letters;* on the contrary, it compounded insult with injury and defended them to the hilt. There was now no alternative but for the Assembly of the Clergy to attack the Jesuits even more fiercely than before: the *Defence of the Casuists* must be condemned, both by the ecclesiastical authorities and by the highest of French courts, the Parlement de Paris. With this end in view, they decided to compose a *Factum,* or official remonstrance setting down the facts of the case, to be laid before the competent authorities. For help in the composition of this they turned to the Jansenists of Port-Royal des

Champs; and the Solitaries, in their turn, entrusted Pascal with the task. Thus, around 25 January 1658, the first *Deposition on Behalf of the Parish Priests of Paris* came to be written.

The seriousness of its tone stems principally from the circumstance that this was the official remonstrance against probabilism made by the whole body of the Parisian clergy. And the same was to be true of the numerous *Depositions* which came later. In all his writings against casuistry Pascal wrote either pseudonymously or anonymously. But the pseudonymous work was a work of satirical fantasy and great imaginative power. Twenty years or so before *The Princess of Cleves* was written, it came close to the status and quality of a novel. Whereas the *Deposition on Behalf of the Parish Priests of Paris* was a quasi-legalistic enumeration of the charges on which a condemnation was sought. This, however, did not prevent it from also being in its own right a work of great forensic and imaginative – but not fictional – power.

Pascal opens the attack squarely: 'Our cause is the cause of Christian morality. Our adversaries are the casuists who are corrupting it' (906). These casuists are gaining in boldness and impudence day by day, to such an extent that they are now not only the corrupters of public morals but also the corrupters of the very laws which are the ultimate sanction of good and evil within society. This, in his view, is a far more heinous and corrosive sin; for it is one thing to induce a certain laxity in moral standards, but quite another to undermine the laws themselves: 'once that barrier has been removed, concupiscence spreads unhindered, so that there is no difference at all between making vices permissible and making all men vicious' (907). These laws, Pascal emphasizes, are fundamentally holy laws; whereas the basis of true morality is the authority of God, whilst its sustaining purpose is love for one's fellow men, the Jesuits on the other hand have substituted reason for divine authority and have replaced altruistic charity by the natural unbridled passions of man. And this brings Pascal back to his essential and characteristic charge: that 'instead of adjusting men's lives to the precepts of Jesus Christ, these new theologians have undertaken to adjust Jesus Christ's precepts and rules to the interests, passions and pleasures of men'.

Scoring every point against them that can fairly and honestly be made, he now spells out some of their favourite maxims (908): that the opinion of one grave author suffices to make an opinion probable; that it is permissible to kill anyone who harms or insults us; that duelling is allowed; and simony likewise. This new morality, he scornfully concludes, gives its blessing to human inclination rather than to the will of God – extinguishing, in fact, men's love of God and the divine will by declaring that men need not love Him provided they do not hate Him.

This ferocious charge is followed by a historical account of the rise of casuistry. Pascal points out how the Jesuits, first abroad, and then in France, gained in self-confidence by gradual stages to the point where they finally felt strong enough to declare quite publicly that any probable opinion was as good as any other. Nevertheless, the Church did not submit quietly to this indignity: the new school of thought was condemned first by the Assembly of the Clergy in 1642, then by the Sorbonne, then by the Theological Faculty of the University of Louvain, and finally by the Archbishop of Paris himself. 'We, by virtue of our knowledge of the innermost workings of men's consciences, noted the harm which was being caused by these disorders' (910), particularly through the doctrines of probabilism (911), direction of intention, and attrition.

Tactically speaking, Pascal has indulged in so long a preamble in order to give devastating force to that final indignity of all indignities, Pirot's book. With 'quite extraordinary boldness' (912), he maintains, Pirot and his friends now reaffirm their belief in their various odious doctrines, openly arguing that every single one of them is justified. Thus (912–13), in their opinion, usury is no longer a sin,[1] nor is it a sin to break the Ten Commandments;[2] menservants may steal from their masters up to the amount they consider their wages ought to be,[3] judges do not have to hand back any bribe that may have caused them to pronounce an unjust judgment,[4] and a man may Christianly partake of the sacraments feeling no other regret for his wrongdoing than because of the immediately temporal harm that wrongdoing may have done him within the world.[5] But Pascal believes that the one inevitable outcome of these tendencies is a breakdown of all traditional moral values. He has virulent scorn (913) for the view[6] that in the killing of our neighbour the promptings of natural reason are a sure guide. He appeals therefore both to the public and to judges, urging the former to ignore Pirot and the latter to punish those who are influenced by him to do evil.

What Pascal finds most alarming and surprising is the open encouragement given by the Jesuits to Pirot's book. At the Jesuit Collège de Clermont in Paris it is freely on sale. The Jesuits have been spreading it far and wide across Paris and the provinces. At

Rouen it has been read during mealtimes in their College refectory. They have sought permission for it to be reprinted, and have tried to get it approved at the Sorbonne. Only in the parishes of France has there been a quiet but unflagging attempt to resist it. Worse still, says Pascal (915), the parish priests have been vilified by the Jesuits on account of their protests to the Assembly of the Clergy more than a year before. Accusations that they were 'ignorant, factious and heretical, wolves and false shepherds of their flocks'[7] have sown discord within the Church. Yet those who are so unjustly slandered as wolves in sheep's clothing believe that you should never slander your neighbour, always turn the other cheek, and always consider duelling a sin and a crime.

His strongest censure is reserved, yet again, for the notorious remark of the casuists that natural reason will enlighten us whether or not we may kill our neighbour. Not even in pagandom, says Pascal, would such a morality be acceptable! '*Woe unto us if we do not preach the Gospel;*[8] and woe unto us again, say these men, if we do preach it' (916). In a typically Pascalian paradox, he drives the priests into facing up to the nature of their dilemma: either they will be wolves and false shepherds of their flocks or else they will be 'lacerated as such by thirty thousand mouths joined in outcry against us'. *It is time for the Lord to act, for Thy law has been broken. Therefore I love Thy commandments above gold, above fine gold. Therefore I direct my steps by all Thy precepts; I hate every false way.*[9] It is a deplorable thing to have to fight both the fiercely engrained passions of men and all the authority of a great religious order – and to rely on what still remains of dignity and morality within the flock in order to correct the erring ways of many of their shepherds! Pascal concludes this admirable philippic with a reference to the Bull of Innocent X condemning the Five Propositions. This Bull, Pirot claims, signifies 'a general approval of the casuists' doctrine' (917), whereas to Pascal all that probabilism and the Five Propositions have in common is that both are heretical. Yet probabilism is still more insidious than the Propositions; for whereas the latter are fully accepted as heretical by all concerned, and are besides an abstruse matter only for theologians, the doctrine of probability is a heresy publicly upheld by the Jesuits and directly concerning everyone.

If the first *Deposition on Behalf of the Parish Priests of Paris* is an admirable philippic, beautifully balanced, nicely ironical and supreme in its handling of paradox, the *Second Statement by the*

Parish Priests . . . is perhaps still more admirable, in its marshalling of its material, dignified self-control and talent for carefully introducing new perspectives whilst never losing its hold on the public's interest. It was issued on 2 April 1658, in reply to a protest from the Jesuits that they had been scandalously misrepresented.[10] The Jesuits' reply to the *Deposition* cunningly claimed that it did not seem to have been written by the parish priests of Paris at all, but had been issued mendaciously in their name. Pascal deals first of all with this calculated ambiguity of their attitude towards the parish clergy, then widens the discussion to include the bishops, emphasizes yet again the most pernicious aspects of the casuists' doctrine, and finally comes to perhaps the most sensitive subject of all: the accusation that they, the parish clergy, are spreading strife, division and schism within the Church. In dealing with the baseness of this final accusation, Pascal's sustained eloquence rises to unparalleled heights. Said to have been written in a single day (905), this *Second Statement by the Parish Priests of Paris* has all the dynamic verve and irresistible fluency of works kindled into incandescence by a towering anger and so composed at a single sitting. Perhaps the most brilliant of all examples of this genre, it is also Pascal's finest piece of completed connected writing, though many passages both from the *Provincial Letters* and the *Thoughts* stand equal or superior to it as isolated fragments.

After expressing the hope that the Jesuits would now have kept silent, Pascal points out that the pamphlet just published by his adversaries (918) not only reasserts all the disputed points of casuistry but tears into the *Deposition* in a most outrageous manner; so that the parish priests have no choice but to reply. This *Deposition*, say the Jesuits, is a fabrication not written by the clergy of Paris (in a limited sense, of course, the Jesuits were right!). The clergy therefore reaffirm that it was their document, published with their authority. Pascal then shows, with delicious irony, how his opponents attack the *Deposition* but extol its 'supposed' authors, as if it were some vile imposture unworthy of Christian ministers. Their high-flown and extravagant praise of the Parisian clergy contrasts strangely with all that had been said about them in their *Defence of the Casuists*. Whereas in Pirot's book the parish priests had been denounced as 'false shepherds', 'ravening wolves', 'ignorant', 'heretical and schismatical' (919), now on the contrary they are commended as 'true and worthy shepherds' etc., etc. However, between both cases there is one important point of resemblance: the fact that both of the Jesuits' publications uphold a corrupt morality. Thus, 'the change in their style is not a consequence of the conversion of their hearts, but is a skilful political manœuvre'. They remain the enemies of truth and of truth's defenders. 'All the harsh things which they appear not to say about us as parish priests, they say about us as authors' (920). What is more, the Jesuits' strictures also apply to the circular letter sent by the bishops of the Assembly of the Clergy to their fellow bishops throughout France. This too, they claim, is a spurious and mischievous invention; but again Pascal insists on its genuineness and authority.

Thus the argument is led to the point where the Jesuits honour the bishops but do not accept their *Letter*, honour the clergy but do not accept their *Deposition*, 'honour the ministers of the Church as long as they do not disturb them in their disorders' (921): only because of their powerful worldly backing can they make such play with the truth of the Church. The by now familiar charges are laid against the casuists yet again: that one may kill one's neighbour on the strength of an impulse of natural reason; practise simony; steal from one's employer, etc., etc. This time, however, the mention of simony enables Pascal to shoot one of his deadliest shafts – for, he says, it is precisely through simony that the wolves will get into the fold to devour the flock! All these tenets of casuistry are beyond anything permitted by 'Jews, heathens, Mohammedans and barbarians' (922).

The most distinctive argument, however, of the *Second Statement* is that concerning the propagation of schism and discord. It seems that this was a point on which Pascal had frequently to defend himself against some of his friends who, in matters of what they regarded as religious truth, were all too content to turn the other cheek. On whom lay the onus for perpetuating the strife and discord? The Jesuits would have blamed the parish priests except that they 'know' that the clergy would not wish to disturb the peace of the Church; Pascal lays the blame fairly and squarely upon the Jesuits.

> We have only spoken when it would have become criminal of us to remain silent . . . The true children of the Church well know how to distinguish that true peace, which the Saviour alone can give and which is unknown to the world, from that false peace which the world can indeed give but which the Saviour of the world holds in abhorrence (923).

When, therefore, the truthfulness of the Church is challenged, what men think of as peace is war in the sight of God. Seeking to 'establish their human traditions amidst the ruins of the divine ones', what the Jesuits merit is *'not . . . peace but a sword'*.[11] *'There is a time for peace and a time for war'*,[12] writes Pascal, but never 'a time for truth and a time for untruth' (924). St Athanasius, Hilarion and many others were called in their time 'rebellious, factious, stubborn, the enemies of peace and unity' (925). At the irregular Council of the 338 Iconoclast bishops held at Constantinople in 754 St Stephen the Younger was accused of disturbing the peace of the Church – and, after ten years of unsuccessful attempts to seduce him into heresy, put to death – because he condemned the fanatical campaign against sacred images led by the Byzantine emperor Constantine V Copronymus. Nevertheless, 'one should never abandon the slightest articles of truth on the pretext of preserving harmony'. Even the Apostles themselves disturbed the peace of the Church by castigating the Pharisees (925), just as the Parisian clergy are now disturbing the 'Pharisees of the New Law' (923).

Pascal even appeals to the authority of the Wisdom of Solomon XIV 22 in his claim that the 'great war of ignorance' (*magno . . . inscientiæ bello*) can sometimes be foolishly, and paradoxically, mistaken for peace. He arraigns the Jesuits: 'We see before us the most numerous and powerful Company of the Church, which directs the consciences of almost all men of great rank, leagued together and bent upon defending the most terrible maxims beneath which the Church has ever groaned' (926). Must the clergy keep silent, when to save both themselves and the whole Church they should cry to their leaders: *Save us, or we perish!*.[13]

> The least truths of religion have been defended unto death, yet we are asked to relax the most essential points of our religion – those maxims which are the most important and most necessary for salvation – because it is the good pleasure, not of 300 bishops, nor even of one, nor of the Pope, but merely of the Society of Jesus, that they should be overturned.

If the Jesuits will drop their pernicious maxims, the clergy will offer friendship and reconciliation. They emphasize that they are in no way motivated by reasons of petty spite. The same protests against casuistry are being made at Rouen and in other French cities. *'We shall be turned aside neither by their maledictions nor by their benedictions,* in the words of the Scriptures.[14] They have not intimidated us by their enmity, nor will they corrupt us by their flattery . . . We shall only resist the duplicity of the *children of this world*[15] with the simplicity of children of the Gospel' (927–8).

Great as are the qualities of the *Second Statement*, in Pascal's own view the *Fifth Statement by the Parish Priests of Paris*[16] was 'the finest thing he ever wrote'.[17] He now considers the Jesuits' innovations from the standpoint of those who are outside the Roman Catholic Church, in particular the Calvinists. The difficulties created by the *Defence of the Casuists* are of a different order, but just as great outside the Church as within it. For, to justify their heretical separation from the one true Church, the Calvinists are now able to point to the *Defence* as an illustration of the untruth of the Church and the degradation into which it has sunk.

'*Licence*', they are able to say (928), 'is rampant everywhere'; and, to prove his allegation that the true Church is being betrayed by the Jesuits' antics, Pascal quotes from the writings of the Protestant minister Charles Drelincourt to the effect that a man need not love his Creator, whilst another minister (unnamed) is said to have quoted the teachings about killing one's neighbour, judges repaying bribes, etc. as the ultimate in moral depravity. For years these accusations have been going on, but now with the appearance of the *Defence of the Casuists* they are more frequent and justifiable.

Worst of all, these corrupt doctrines emanate from a respectable body within the Church (930), so that whenever the Calvinists blame the Church for holding these views and the Church rises up in self-defence, the Jesuits defend themselves too and further dissension within the Church is sparked off. Yet, writes Pascal, the clergy of Paris know that the Jesuits and the Calvinists are not really alike and that the former are really the latter's enemies. The Jesuits 'would wish the world to be inclined towards the severity of the Gospel, which they only corrupt in order to adjust it to human nature'. Nevertheless, the Devil uses both sides against the one true Church: each fortifies the other in claiming that the Church really believes these things. If God were to permit the outcome towards which both are working, all men would be damned. Yet the fact that Jesuits and Calvinists are allies makes the Church's self-defence all the easier, for the more lies are heaped up against the truth, the more clearly does the truth stand out (931).

The guarantee of Christian truth is, in Pascal's view, tradition: the legacy of divine truth handed on in an unbroken Apostolic

Succession from Jesus and His disciples. The pernicious maxims of casuistry have never been held either by the Fathers or the Councils of the Church. The Jesuits have lied about this, but their trumped-up quotations from the early Fathers have been exposed in Arnauld's *Third Statement*. The very strength of the protests of the French clergy and Catholic faculties proves that they have not broken faith with their tradition, whilst the separate excesses of Calvinism and the Jesuits prove that that true tradition should never be ignored. The Jesuits are destroying the salvation of many, both within and outside the Church: within, because many are turned aside from their Catholic belief by such perversity; outside it, because heretics are equally repelled by it. Only two solutions remain: either to reform or to denounce the Society of Jesus. The first depends on themselves, and they will not do it. The second depends on the parochial clergy, and they must do it.

Pascal concludes his *Fifth Statement* on what he regards as an important doctrinal issue. Although both Jesuits and Calvinists are misguided, the latter are so much more misguided than the former. At least the Jesuits are 'within the unity of the Church' (936), whereas heretics are enemies outside the fold. Schism is always to be avoided. 'There is absolutely no excuse for the Calvinists', writes Pascal, because (quoting St Augustine)[18] *'breaking away from the unity of the Church is never truly justified'*. Because the Jesuits are corrupt, it does not follow that the Calvinists should not be within the Church. We after all, the parish clergy of the archdiocese of Paris, are within it! The Jesuits' one virtue is that they have preserved unity; no doubt a few good Jesuits exist, but there can be no good Calvinists since all are heretical. Moreover, sacrifices and prayers can still be offered up for the Jesuits, as they are within the unity of the Church. This is not so of the Calvinists, and therefore Pascal is led to his ultimate paradox: 'it can truly be said that the heretics are in so unfortunate a condition that, for their own welfare, one might wish them to be similar to the Jesuits' (938). Nevertheless, the ordinary Catholic congregation has more to fear from the Jesuits than from the Calvinists, precisely because the Jesuits are of their own number. Their perverted casuistry is 'a much lesser evil than schism', yet more dangerous because closer to the natural dictates of the heart. 'Even though they are members of our body, they are sick members and we must avoid their contagion'.

Whatever Pascal himself may have thought of the merits of the *Fifth Statement*, there is perhaps a rigorism about it – a love of categories, an excess of paradox, a tendency to see issues as unreservedly clear-cut – which prevents it from being considered his finest work. The fact that he thought of it as his most distinguished literary achievement tells us something also about its author. Some may accuse Pascal of a lack of charity towards those other denominations of the Christian Church which are nowadays frequently referred to as the 'separated brethren'. For all his intellectual subtlety, he has as much a tendency as his contemporaries to view religious questions in black and white. The Jesuits are 'falsifiers' (933) if not 'forgers' (the word is the same); all heretics are enemies; no heretic is free from error; no heretic can be in charity with God since all have transgressed His unity (937). Of all his writings on religious matters, the *Fifth Statement* reveals Pascal's mind at its most mathematical – with all that that implies in the way of argument from fixed premises, and preference for a *priori* reasoning. It is not known when in his life Pascal considered the *Fifth Statement* to be his finest writing; but it is doubtful whether he would still have done so as he became more and more immersed in the complexities and nuances of the *Thoughts*. Pascal, however, was not lacking in charity towards his 'separated brethren', the Calvinists. It was simply that, in line with the thinking of the Counter-Reformation, his conception of charity was dogmatic and restrictive. Out of charity for his 'separated brethren' he sought to win them away from a religious adherence which he regarded as lethal to their salvation in the Hereafter: the *Fifth Statement* makes it clear (on the assumption that he, not a member of the clergy, wrote these lines: **VII 371**) that he accepted the doctrine of divine retribution in an afterlife, and 'fire and brimstone' for those whom God finally condemns. But whether or not every line in the *Fifth Statement* is by Pascal, there is a density of texture, and an elaborately interwoven quality illuminated by clarity of intellectual outline, which are certainly his. It also has something of the flashing brilliance of a juggler, and is a little prolix and repetitive.

The *Fifth Statement* was issued by the clergy of the archdiocese of Paris on 11 June 1658. Other depositions and petitions both preceded and followed it, some of which Pascal wrote and in all of which he may have had a hand. At least two are unquestionably his, the *Draft Mandamus Against the 'Defence of the Casuists'* and the *Sixth Statement by the Parish Priests of Paris*. Meanwhile, a *Deposition on Behalf of the Parish Priests of Rouen* – possibly but not certainly by Pascal – had appeared on 15 February 1658; a *Petition by the Parish*

Priests of Amiens and Nevers (5 July), a *Petition by the Parish Priests of Beauvais* (10 July) and a *Deposition on Behalf of the Parish Priests of Amiens* (27 July) are also amongst the writings sometimes attributed to Pascal, whilst the *Deposition on Behalf of the Parish Priests of Nevers* (27 July 1658) is almost certainly by him. All these depositions and petitions were requests from the diocesan clergy to their respective bishops to condemn the *Defence of the Casuists*. The *Draft Mandamus*, which was never published and does not seem to have been used during Pascal's lifetime, was no doubt drawn up for the use of one of the bishops or archbishops (perhaps François de Harlay, Archbishop of Rouen); certain parts of it still exist in Pascal's manuscript, whilst from internal evidence it is fairly safe to assume that it was composed around the time of the *Fifth Statement* (11 June 1658). The *Sixth Statement by the Parish Priests of Paris*, issued on 24 July, derided the Jesuits for their pretence of neutrality about the *Defence of the Casuists*, when they themselves had initiated the book and so could not disown it. Numerous bishops condemned Pirot's notorious polemic, as did the Sorbonne, other theological faculties, the Jesuits themselves and ultimately, in 1659, Pope Alexander VII.

If the *Depositions* are the prolongation of one aspect of the *Provincial Letters* (the dispute about casuistry), then the *Writings on Grace* are the continuation of the other. The mystery of the nature and operation of grace is a question to which Pascal has made a more eirenic and discerningly speculative contribution. Above all, he is preoccupied by the aspect of divine foreknowledge. How can God's giving or withholding of His grace be reconciled with man's freewill? How can an all-good and all-powerful God effectively foreordain the damnation of many? How can those favoured by Him know they are of the Elect? For if 'natural powers', in the phrase held up to scorn in the fifth *Provincial Letter* (707; 77), are not sufficient for a man to work out his own salvation, then that salvation must depend on the supernatural power – or the arbitrary pleasure – of God. *Thou hast created all things, and for Thy pleasure they are and were created.*[19] But the divine displeasure is then akin to a cruel and omnipotent persecution:

> *As flies to wanton boys, are we to gods,*
> *They kill us for their sport.*[20]

Is not grace, being of divine origin, irresistible? If God's grace is freely bestowed on all men, must not all men be saved? Why indeed, if the semi-Pelagian or Molinist standpoint is accepted, has God willed that some men may accept and others reject His will?

Pascal bases his position on three essential principles (948): the belief that some men are saved and some damned; that those who are saved wish to be saved, just as God wishes them to be (for 'He Who made us without ourselves cannot save us without ourselves'); and that those who are damned likewise wish to merit their damnation, just as God wishes to mete it out to them. The will of man and the will of God are both involved, though perhaps not equally involved, in the events of the Hereafter. Moreover, these suppositions are not pious flights of fancy on Pascal's part but conclusions drawn from a careful reading of the Church Fathers and a pondering of Christian tradition.

Which, however, is the predominant will conducing to salvation and damnation: God's or man's? One of the most distinctive features of the *Writings on Grace* is their eirenic, reconciling attitude to this crucial question. Pascal's approach now embraces Molinism, recognizing that it teaches an important but limited truth. The right use of a man's freewill is an indispensable ingredient in the working-out of his salvation; but it is still only the secondary cause of a man's salvation: the primary cause is the will of God. This syncretism presents both Molinism and Jansenism as interdependent elements in a global, unitary truth emanating from St Augustine.

What the *Writings on Grace* also strongly emphasize is that Jansenism and Calvinism are by no means indistinguishable (969–70, 951). Yet, in keeping with Pascal's tactical method, they never refer to Jansen as such but always to St Augustine: 'Calvin', it is even claimed (969), 'has no point of resemblance with St Augustine, differing from him all the way through in every respect'.

> Those to whom . . . grace has once been given are infallibly saved, not through their good works or good will, for they have neither, but through the merits of Jesus Christ extended to them. And those to whom . . . grace has not been given are infallibly damned on account of the sins which they commit by the order and decree of God Who inclines them towards sin for His own glory (970).

This, to Pascal, is the essential message of Calvinism. Augustinianism, on the other hand, teaches that God has an absolute will to save those who will be saved and a conditional will to damn those who will be damned: for those who are saved would not have been so but for the will of God, whilst those who are damned are cast out by God not through His absolute will and foresight existing from the beginning of time but because they are foreordained by Him for damnation on account of original sin. Original sin stems from the condition which Adam, and through him all men, failed to fulfil in God's plan for the salvation of the world. In the beginning of things this plan was that all men should be saved, and all received *sufficient grace* to enable them, with freewill, to work out their salvation (965–7). But the fall of man meant that Adam and his descendants were all worthy of punishment. It is therefore the conditional will of God that some, but not all, of these shall receive their punishment (952–4).

Those whom God elects for salvation receive His efficacious grace, in addition to which however a *proximate power* is necessary which will enable them to persist in prayer (976–7). This proximate power (a reminder of the somewhat arid debate between Dominicans and Jesuits in the first *Provincial Letter*) is even sometimes withheld from the Elect, however – for Pascal establishes (976), in a tightly argued syllogism, that

> i) if all the Elect have a proximate power to pray 'in the next instant', they must also have a proximate power to persist in prayer;
> ii) the Council[21] has laid down that not all the Elect are capable, at all times, of such persistence.

Therefore

> iii) it is contrary to the teaching of the Council [of Trent] to say that a proximate power to pray 'in the next instant' is always given to the Elect.

The upshot of such theological nicety is that *God sometimes deserts the Elect before they desert Him*, for the efficacious grace needed for salvation comes to man through prayer, but the Elect may not always be able to pray for this grace 'in the next instant': God, therefore, in the aftermath of the fall of Adam, chooses the Elect and chooses the times when the Elect are not given that efficacious grace on which their salvation ultimately depends. Does this mean, then, that some of the Elect will not be saved? Conversely, may salvation finally be given to some of those for whom God foresees and foresaw no salvation? Pascal does admit (953) that the non-Elect are sometimes 'called to the condition of the Elect, thus participating in the Redemption of Jesus Christ: it is their fault if they do not persevere; they could if they so wished; but not being of the Elect, God does not give them the efficacious grace without which they never actually can do so'. As for the Elect themselves, their salvation is assured: they have that perseverance which alone enables one to remain in the faith, and which is God-given through Jesus Christ (954).

The key to salvation is prayer, the golden formula by means of which the Elect will remain for ever at one with God. The latter half of the *Writings on Grace*, once the hard clear lines of the theology have been marked out, is a sort of vibrant and tremulous meditation on the ways in which a man may be abandoned by his Maker. For this abandonment no apparent reason exists (984–5, 966, 987, 953) beyond the fact that God in His omniscience can foresee the sins which will be committed by the freewill of man (951, 953, 965). Prayer is the mediation of grace (996), and this grace may be considered either efficacious or sufficient (968): 'continuing instant'[22] prayer being *sufficient* to procure grace which will be *efficacious* to procure salvation. Nevertheless, that prayer is still dependent on the will of God that one should pray – and God can still abandon the prayerful man:

> It is true that, for the performance of good works, God never withholds His help from those who ceaselessly ask it of Him, and in this sense God does not abandon the Justified man until the Justified man has abandoned Him; but it is also true that God does not always give His help in praying, and in this sense God abandons the Justified man before the Justified man abandons Him: the result being that this abandonment is always so arranged that, first, God leaves man without the help necessary for prayer and that, then, man ceases to pray and, then, God leaves the man who no longer prays to Him (991).

Thus, in a dizzying plunge into the abyss, Pascal even conceives of a double abandonment of the Elect by God. A related question, which in 1547 had been answered in the affirmative at the Council of Trent[23], provoked Pascal's most earnest and anguished speculation: *Are God's commandments always performable by the Elect?* (970 *seq*) This seems to have filled him with extreme disquiet not only because of his general respect for the Tridentine doctrines (as indeed for all the pronouncements of ecumenical councils) but because it again struck at a peculiarly responsive chord within his own psychology. He studies it from every angle, quotes at length from St Augustine, St Fulgentius, Prosper of Aquitaine, and the canons of the Council of Trent, contrasts the *equivocal* with the *univocal* elements in the sayings of Jesus (987–8), ponders the ambiguities and apparent contradictions within Augustine's own writings, heaps up the arguments in favour of Church tradition, turns again to the Gospel, and, notwithstanding the Council of Trent, can only conclude that 'some commandments are not sometimes possible for some of the Elect to accomplish' (988, 977): yet, he asserts, there is no inherent contradiction in this, because though 'God never refuses what is actually asked of Him in prayer' (977), nevertheless He does not always give men the effectual power to ask – and so we return to the heart of the enigma.[24] On one point, however, Pascal is firm, clear and utterly convincing both to himself and his reader (995–6, 954); and it is a point of substance. How, in a non-Pelagian world where freewill, good works and the voluntary turning of the human heart towards God are apparently of no effect, can any incentive remain for human effort? The world of St Augustine (and therefore, in a sense, of Pascal) is a chill baffling universe in which there is no quickening motive for any man or woman to pull themselves out of their lethargy and complacent worldliness, thus striving to redeem themselves and work out their salvation by their own actions. Where God has foreordained and foreseen both the justification of the Elect and the damnation of all the remainder of mankind, what distinctively human hope remains? Pascal is positive in his affirmation that, at least on a terrestrial plane (which is all by which we can judge), all men must act *as if they were of the Elect*: all, he writes,

> must believe, but with a belief mingled with fear and unaccompanied by any certainty, that they are of that small number of the Elect whom Jesus Christ wishes to save, and never assume that any man living upon earth (however wicked and impious he may be, so long as a moment's life is left in his body) is not of the number of the Predestinated, leaving to the impenetrable secrecy of God the separation of the Elect from the damned. Which means that they must do for themselves [everything] that can contribute to their salvation (954).

The judgment in the Hereafter is mystery, aloof and impenetrable; but the duty on this earth is a certainty, clear and straightforward, a kind of divine Categorical Imperative bidding men to act as if the mainspring of their action were the Justification of the righteous. Otherwise, however, all is unfathomable, and the tone even a little presumptuous – as if the Council of Trent, and all those basing themselves on the accretions of Church tradition, ventured to know all the intricacies of the workings of God's grace in its inspiration of men.

> If we find that it is a firm principle in St Augustine that all those who now have prayer have it through an efficacious grace, and that none of those who do not now have prayer does have the proximate power to pray, will not the question be resolved, and will it not necessarily follow that whenever the Elect pray, they are sustained efficaciously, and that they never cease to pray so long as this efficacious help is by them, and that when they cease they do not have the proximate power to pray? And consequently that God has left them, and not they Him: I do not say without any help, but without proximate help . . . (996).

For once even Pascal seems almost at a loss how to continue. The reasonings become more tortuous, eliminating the awesomeness of the mystery, the tempo quickens . . . and Pascal is left facing the very depths of the theological abyss, beyond which no further journey of the human mind is possible, no positive answer to the unknowable remotely reassuring. The finite having (in Dryden's phrase)[25] failed to grasp Infinity, Pascal now turned from theodicy to two areas where finite reason could grasp and achieve results of incalculable importance, proportionate to the human condition. He was never to tackle a treatise on divine grace again.

A bout of toothache, one sleepless night in June 1658, not only inspired his discoveries about the cycloid but – according to his sister Gilberte (19–20) – was the actual starting-point of the desperate worsening of his physical condition which from now on would be continuous, leading by slow but unrelenting degrees to his times of utter prostration and eventual death just over four years later. The cycloid research, she writes, 'was not too much for his mind; but his body could not withstand it, for it was this final affliction which eventually resulted in the complete undermining of his health, reducing him to that very sorry condition . . . in which he was unable to swallow' (20). And not only did his illness prevent him from continuing with his mathematics, it was also to gravely handicap his writing about religion. In Gilberte's words: 'his infirmities made him incapable of serving others, yet nevertheless they were not without value to [the patient] himself'. She sees her brother's immense afflictions as a kind of purification wrought by God so as to enable Pascal to appear pure and spotless before Him at the last day. And indeed, she adds, 'henceforth he thought of nothing else; and, having always before his eyes the two maxims he had set himself of renouncing all pleasures and superfluities, he now observed them with even greater fervour, as if propelled forward by the impetus of feeling that he was drawing near to the centre where he would enjoy eternal rest'.

Probably in the first half of 1659,[26] and perhaps at the time (in June of that year) when he was reduced to drinking asses' milk (**IX 202 n. 1**), Pascal composed his *Prayer to God Concerning the Proper Use of Illnesses*. He taught it to his sister, for her own use, not as an outward observance, a sort of talismanic device for the banishment of illness, but as a supreme self-offering to God in the confidence that illness has something positive and valuable to teach mankind. In Gilberte's eyes Pascal provided the matchless example of Christian witness at a uniquely sustained level: 'he wrote in this way about these things simply because that was how he lived. We can even give an assurance that we ourselves have been witnesses of this . . . No one has ever surpassed him in that exercise, with greater edification for all who beheld him'. The *Prayer to God Concerning the Proper Use of Illnesses* lays its emphasis on the God-given nature of sickness and indeed of all suffering. Illness is not an irksome nuisance to be avoided whenever possible and recovered from at all costs; *not as I will, but as Thou wilt*[27] is the prayer's moving spirit: sickness and suffering are forms of purgatory or purification on earth and as such are to be rejoiced in by the truly devout believer.

> O Lord [the prayer begins], Whose spirit is so good and so tender in all things, and Who art so merciful, so that not only the prosperities but even also the afflictions which are the lot of Thine elect proceed equally from Thy mercy: grant of Thy grace that I may not act as a pagan in the condition to which Thou in Thy justice hast reduced me; and that like a true Christian I may acknowledge Thee as my Father and my God, whatever the condition in which I may find myself . . . ; [recognizing] that Thou

> art the same, though I am subject to change, and that Thou art no less God when Thou dost punish and afflict me than when Thou metest out to me leniency and consolation . . . (605–6).

Thus, the first characteristic of illness, and especially of prolonged illness, is that it is a trial of the steadfastness of the sufferer's belief in God. God in His majesty may afflict or console His human creation according to His good pleasure. The fact that so much of our human suffering and sickness can be considered the consequences of our voluntary and sinful misuse of freewill is not of any particular importance to Pascal, who does not essentially believe in God's gift of freewill to men (though all men, the righteous and unrighteous alike, must indeed act *as if* endowed with freewill): just as God in His omniscience can foresee our sins, so He in His providence distributes our mortal allotment of joy and sorrow. Nevertheless, in a way which is beyond human understanding, the pains of physical illness are 'both the symbol and the actual punishment of the soul's disorders' (609), as if the sins committed by us were voluntary acts and therefore we, who commit them, deserving in ourselves of condemnation. 'Touch my heart with repentance for my faults, since without such inward pain the outward ills with which Thou afflictest my body would be no more to me than a new occasion for sin'. But the sickness of the body has a much more positive function than merely symbolizing and punishing a spiritual malady. It must also be the means whereby the soul, 'quite sick and all covered in ulcers', is purged of that malady – a gateway to the spiritual self-awareness without which no salvation is possible. For to Pascal the greatest of all sicknesses of the soul is 'that insensibility and extreme weakness which robbed it of all feeling for its own wretchedness':

> Grant that I may feel this keenly; and let my remaining days be a continual penitence, so that I may cleanse my life of the offences I have committed.

Bodily sickness is the way to spiritual health; and perhaps only so can we attain perfect purification and true humility. Opening with an invocation of God the Father, the *Prayer to God Concerning the Proper Use of Illnesses* concludes with a poignant supplication to Jesus Christ. God the Father is the Sovereign Judge before Whom Pascal must 'render a faithful account of [his] life at [his] life's end, and at the end of the world' (606); He 'only allows the world and all the things therein to subsist in order to test [His] elect and to punish sinners'; He 'permits sinners to remain hardened in the delightful and criminal usage of the world, and of worldly pleasures'; He will 'consume at the last day both Heaven and earth and all the creatures therein contained, in order to show to all men that nothing is immanent except [Himself], and therefore that nothing is worthy of love except [Himself], since nothing is lasting except [Him]'. Pascal, believing in the eschatological judgments of the Last Day, praises Him for His goodness in detaching his soul, through illness, from the snares and passions of the world:

> I praise Thee, my God, and shall bless Thee all the days of my life, because it hath pleased Thee to reduce me, so making me incapable of enjoying the sweet delights of good health and the pleasures of the world; and because Thou hast, as it were, destroyed to my advantage the deceptive idols which Thou wilt effectually destroy to the confusion of sinners on the day of Thy wrath . . . (606–7).
>
> Grant, that I may regard this illness as a kind of death, separated as I am from the world, deprived of all the objects of my attachments, and alone in Thy presence to implore of Thy mercy the conversion of my heart . . . (607).

It is not, therefore, enough that a severe and prolonged illness should have detached Pascal's soul from the deceptive idols of worldly passion, though God be praised for visiting him with such physical affliction. Merely to avoid the commission of sin through happening to lack either the opportunity or the desire to commit it would be too negative an attitude for salvation, a dilution of human and divine responsibility akin to the *attrition* of the Jesuits. Pascal's praise of God for detaching him from the world is accompanied by an entreaty which is at the very core of the *Prayer . . . Concerning the Proper Use of Illnesses*: the plea that out of this detachment there may come, in solitary confrontaton of his Maker, a *contrite* heart.

Mysteriously, however, the solitary confrontation of the Father does not exclude the Son. Jesus is with Pascal in his illness, as He will be with him in his death, and with and in Jesus is the active revivifying power of the Holy Ghost. The poignant supplication with which the *Prayer* ends asks humbly for the mediation of Jesus between Pascal and his Maker, the propitiation through Him of the divine mercy, and for a sharing with Jesus of his own private suffering. The Passion of Jesus will continue throughout all time, through the Church which is His Body and through the community of believers who are the members of that Body. Pascal prays that some part of that sacred Passion may be lived out in his own sickness, and that thus his suffering might be to the greater glory of Jesus just as he and countless others with him are both saved and glorified in the divine plan of redemption:

> Grant therefore, O Lord, that such as I am I may obey Thy will; and that, sick as I am, I may glorify Thee in my sufferings. Without them I cannot attain glory; without them, my Saviour, Thou Thyself wouldst not have attained it. It was by the marks of Thy sufferings that Thou wast recognized by Thy disciples; and it is also through sufferings that Thou dost recognize those who are Thy disciples. Acknowledge me therefore as Thine in the ills which I endure, both in my body and mind, for the offences which I have committed: And because nothing is acceptable to God unless Thou offerest it to Him, unite my will with Thine, and my pains with those which Thou hast suffered: grant that mine may become Thine. Unite me with Thee, and fill me with Thee and Thy Holy Spirit. Enter my heart and soul, there to suffer my sufferings, continuing to endure through me what remains to be suffered of Thy Passion as lived out by Thee in Thy limbs until the perfect consummation of Thy Body; so that it is no longer I who live and suffer but Thou Who livest and sufferest within me, O my Saviour: and that as I thus have some small part in Thy sufferings, Thou mayst entirely fill me with the glory which they have won Thee; in which glory Thou livest with the Father and the Holy Spirit, throughout all ages, world without end. Amen (613–14).

Readers of Paul Claudel[28] will recognize much the same sentiment in the fervent wish of the spiritual heroine Violaine to burn out, in her leprosied dying body, some small portion of the pain and evil in the world; Jesus's suffering shared gladly by His creation until the consummation of all things; redemption on a cosmic scale, not 'once, only once, and once for all' but continuing now and throughout time in Jesus's mystical Body: a sentiment which is original neither to Claudel nor Pascal, since it can be found in many of the earlier Christian thinkers and mystics, for example in Thomas à Kempis's

Imitation of Christ. But what is new, at any rate within the totality of Pascal's own work, is the suggestion that perhaps the circle of those whom God chastises in order to save them is wider than the narrow circle of the elect. For what otherwise would be the purpose of the pain and suffering experienced by those who were not of the elect, unless inflicted by a savage and tormenting God? In such a God Pascal cannot believe, and therefore the wider mystery of sickness and physical pain is perhaps to be looked on as a vale of tears and anguish intended for the correction not merely of God's favourites but of mankind generally. Thus, the *Prayer to God Concerning the Proper Use of Illnesses*, though far from offering a reasoned and systematic theodicy, offers an indication (or maybe nothing more than a glancing insight) that the mystery of pain cannot be shirked by the devout sufferer. If it does not diminish one's belief in God's goodness, then it must increase it. A rational theodicy may even be impossible, but the anguish and suffering of the here-and-now, as of the past and also of the future, is an ever-present reality; and from such private and intensely felt contact with a reality physical in that it affects the body, yet also spiritual in that sickness leads away from the body into the unknown, all religious speculation must start.

The three *Discourses on the Worldly Condition of the Great*, perhaps composed little more than a year after the *Prayer*,[29] are the last of Pascal's sustained spiritual exercises. They are, however, spiritual exercises with a difference; for though teaching and illustrating a spiritual truth, they are not intended either for Pascal's own private meditation or for the direct instruction of his fellow-men. Instead, they have the strictly practical purpose of weaning a very young man of high worldly rank away from an over-fond attachment to the world, instructing him both in his duties and responsibilities towards the world but equally in his responsibilities towards himself in relation to the Hereafter. Yet, just as Jesus Christ's own encounter with the rich young man,[30] His parable of the widow's mite,[31] and His saying concerning the camel and the eye of a needle[32] have a wider relevance than to the two ends of the social spectrum to which they strictly refer, so Pascal's *Three Discourses* also have a general application to the human condition. Indeed, like much of the teaching of Jesus regarding worldly wealth, they too are expressed in parable form. As a parable, they embody both a specific and a universal truth. Specifically, they are addressed to Honoré-Charles d'Albert, Marquis d'Albert, eldest son of the Duc de Luynes who was the owner of Vaumurier and the Jansenists' protector; in October 1660 this boy, who was to become the Duc de Chevreuse, a son-in-law of Colbert, a friend of Fénelon, and one of the most distinguished and cultivated men of his generation, was aged fourteen and a pupil of Lancelot at the Petites Écoles of Port-Royal des Champs. Nicole, another of the teachers at the Petites Écoles, has put it on record that 'the instruction of a prince whom one would try to bring up in the way most appropriate to the station to which God has called him, and the aptest to equip him to fulfil all its duties and avoid all its pitfalls'[33] was one of the projects dearest to Pascal's heart – and one also on which he held many views. 'He has often been heard to say', Nicole continues, 'that there was nothing to which he would have a greater desire to contribute if he were so invited, and that he would willingly sacrifice his life for so important a matter'. It cannot presumably have been mere vainglory which led Pascal to wish to emulate the position of Aristotle in relation to Alexander the Great, of Seneca towards Nero or Hobbes towards Charles II. In a world of absolute or near-absolute monarchies he was as keen to instruct a young prince in the ways of virtue as the Jesuits were to train him in worldly wisdom perhaps for their own ends. The only astonishing thing, as Nicole himself remarks, is that Pascal never actually wrote such a manual himself. Not one of the *Thoughts* concerning religion expressly deals with this subject, though 'all are in a sense concerned with it'.[34] The *Discourses on the Worldly Condition of the Great* are Nicole's own record, ten years or so after the event, of the three lessons in the duties of a great nobleman which he heard Pascal give to the young Honoré-Charles d'Albert perhaps during a last visit to the Solitaries and their school in August 1659. As such, they are not actually the literary work of Pascal himself, though they sound like his authentic voice. So vivid was the impression made on Nicole by these lessons at which he was incidentally present that, even after many years, he claims to have remembered them clearly. The secret of their eloquence is no doubt that, over and above the specific purpose for which the lessons were intended, the *Discourses* propound a truth which is of the essence of the whole human condition. Unlike Machiavelli's Prince, whose aim was to train young rulers in the arts and wiles of secular statecraft with all its compromises and hypocrisy, or Fénelon's *Telemachus* which is a refined blend of the secular and the spiritual but with its accent on the world, Pascal's *Discourses on the Worldly Condition of the Great* are the reminder above all that the world's glory is transient and that even a prince is first and foremost a man facing the eschatological problems of the human race like any other.

The first *Discourse* opens with the vivid parable of a stranger shipwrecked on a remote island. The king of this island has recently disappeared, but the stranger bears an uncanny resemblance to him and so the inhabitants of the island take him for the monarch they have mysteriously lost. He is acclaimed as such, and yet knows that he is not really their king: in his own heart he knows of his lowly status, though to the outside world he acts with all the appearance of royal authority.

By an equally accidental stroke of chance, says Pascal, the rich, great and powerful of this world have received their wealth, authority and privileges. 'You have no more right to them, of yourself and of your own nature, than he had: and only thanks to an infinity of chance happenings[35] are you both a duke's son and a dweller upon this earth' (616). The young man's very birth depends upon a marriage, and that marriage upon all others before it. And all those marriages depended upon a multiplicity of accidents and unforeseen coincidences. Likewise, the wealth of the great is a mere contingency, the fortunate but by no means necessary outcome of 'a thousand chances', both in the getting of the money and in its retention. And therefore it is not possible to claim that the accumulation and inheritance of great wealth is justifiable in terms of natural justice. In fact, it is a pure convention 'grounded simply and solely in the will of legislators who may have had their good reasons'; but equally, if these legislators had decided that, on its owner's death, a private fortune should pass to the State, no citizen would have any right to complain. Any legitimate complaint must be founded in natural justice, and there is no principle of natural justice which says that a man should or should not be possessed of a great fortune.

On the other hand, Pascal is at pains to stress that the owner of great, even inherited, wealth is entitled to it as the law stands. In this respect, the rich man differs from his metaphorical counterpart, the king who is not really a king. For the rich man is legally a rich man, whereas the shipwrecked stranger was not legally a king. So far as the latter's possession of the kingdom is concerned, 'God would not authorize this possession, and would oblige him to renounce it, whereas He authorizes yours' (617). The rich man and the king are only alike in that neither's claim to his possessions is based on 'any quality or merit'. In the veiled discreetness of his analogy, we may suppose that Pascal is going as far as he actually dares to point out that the rich man and the genuine king are entirely alike. Thus, the spurious king is not only a metaphor for the condition of legally constituted kingship. Pascal is subtly leading us to conclude that the claims of Louis XIV, Philip IV of Spain, the Holy Roman Emperor Leopold I, the perhaps by now restored Charles II of England and all his other royal contemporaries are no more founded in 'any quality or merit' than are the claims of the rich man; yet all are founded in a Divine Right, though not in natural justice. The conditions of the king and the rich man are founded in a sort of Divine Right because, if God did not approve of them, He would change them: a view akin to Alexander Pope's expression of Optimism, seventy years later, that 'whatever is, is right';[36] God has ordered their estates, but not through any merit of theirs. In the age-old dilemma posed by the diametrical opposites of secular and religious authority, Pascal believes that the legally constituted *status quo* is presumptively the right one: he is therefore close to the teaching of Jesus, that we should 'render . . . unto Cæsar the things which are Cæsar's; and unto God the things that are God's',[37] and he is close to that of St Paul: 'There is no power but of God: the powers that be are ordained of God'.[38] How God will 'oblige' a ruler 'to renounce' the civil power except by human revolution, he does not say: the *Discourses on the Worldly Condition of the Great* are not a treatise in political science. But it is certainly a far-reaching and radical suggestion, though basically more metaphysical than political, that the transmission of private property from one generation to another could be entirely abolished by the 'legislators' – and no one would have the right to feel aggrieved.

This suggestion is metaphysical rather than political because Pascal is always thinking of the individual soul rather than the collective needs and aspirations of a community continuing through time. What the recognition of the arbitrariness of social dispensations amounts to (in his opinion) is clearly stated at the end of the first *Discourse*. A rich and powerful man must always live in the 'twofold consciousness' of his dual status: wealth and power are only his because the community grants and concedes these to him. 'Though to the outside world you behave towards others as your rank dictates, you must acknowledge, more inwardly, but also more truthfully, that by nature you are in no way their superior'. Falsely, the common people may think that 'nobility is real grandeur': the noble man must accept, and may in his actions show that he accepts, that this is not so. Although he must never behave slightingly towards his fellow men, Pascal allows him, but only on a strictly worldly plane, to conceal from them his utter identity with them in natural terms as a human being. The common people, he observes, 'consider men of high rank as being almost of a different nature from other men'. It is permissible not to disillusion them; class distinction may be preserved: but all men are equal in the sight of God. The supposedly 'great' must never be tempted by the trappings of power and luxury into forgetting their essential human frailty.

The second *Discourse* – foreshadowing a distinction enlarged and developed by Rousseau, almost exactly a century later, in his *[Second] Discourse on the Origin of Inequality amongst Men* – distinguishes between two kinds of grandeur: the natural and the institutional. These relate, as it were, to two different orders of being, just as in mathematics there are different orders of numerical power which must strictly be differentiated. Far from being complementary or coterminous, these orders of grandeur are discontinuous. The one, entirely independent of the other, bears no relation to it. A man may be of exalted rank, yet in character nil. Another may be of lowly station, yet morally admirable. The institutional grandeur relates to the secular and the temporal. The natural grandeur relates to the spiritual and the eternal. Institutional grandeur proceeds from arbitrary premises: one country may respect an aristocracy, another may respect commoners; one may practise primogeniture, another ultimogeniture; and there is no particular reason why one thing should be the custom in one place, and the opposite in another. Nevertheless, Pascal repeats, such customs are hallowed by usage; and it would be wrong to flout them.

Natural grandeur, on the other hand, is the very reverse of arbitrary. It has nothing to do with customary usage, but is achieved and conceded by reason of a person's 'real and effective qualities of soul or body', such as 'knowledge, enlightenment, courage, health and strength' (618). To each of these forms of grandeur a different form of respect is appropriate. Institutional grandeur will be entitled to the outward and conventional signs of social deference (and those paying such deference must inwardly acknowledge, says Pascal, that it is right that such deference should be paid); natural grandeur will, on the other hand, spontaneously receive that higher and more inward form of respect which is esteem. 'Kings must be addressed on bended knee: you must remain standing in the chambers of Princes. It is foolish and mean-minded to deny them these observances'. But the corollary is just as clearly stated: 'I do not have to esteem you, because you are a duke; but I do have to salute you' (619). Only if a man is both a duke and an *honnête homme* can he expect to receive both the highest social deference and also moral esteem. To any duke who was no *honnête homme* Pascal would 'render . . . unto Cæsar' – but 'would not fail to have that inward contempt for [him] which [his] low-mindedness would deserve'. 'If as a Duke and Peer you are not content that I should just stand in your presence with my head uncovered, but wish me to esteem you, I would beg you to show me those qualities that deserve my esteem; if you did so, then it would be yours; and I could not rightfully refuse it to you'. But without those inward qualities a man would never gain Pascal's esteem, even if he were 'the greatest Prince on earth'.

From this accent on natural, as opposed to institutional, grandeur Pascal moves on in the third of the *Discourses on the Worldly Condition of the Great* to consider in greater detail the dangers of worldly possessions, and the contrast beween *concupiscence* and charity. *Concupiscence,* from the Latin *concupiscere* (to covet, lust after, or eagerly desire), is a term beloved of Pascal. It means that eager desire for material possessions and sensual enjoyment – including, but not confined to, sexual intercourse – which is what St Paul also means by the word.[39] In brief, *concupiscence* is all that lusting after the world and its beguiling pleasures which Pascal so deeply deplores. Charity, on the other hand, is the Latin *caritas* (or ἀγάπη, as the word is used in the New Testament): love for one's fellow men, pure and undefiled, unclouded by any sexual desire, of which the highest embodiment is Jesus. It is the virtue, higher than all other virtues, whose highest panegyric is given by St Paul.[40] Thus, the antithetical dialectic of Pascal's thought in the third *Discourse* is essentially Pauline.

This dialectic opens with a startling announcement, in a manner which we may surmise would have become more habitual with Pascal, had he lived: 'I wish to reveal to you your true condition, for that is the one thing of which people of your kind are most ignorant'. To be a great nobleman is 'to be the master of a number of things which are the objects of men's concupiscence' (619–20), and thanks to that mastery he will always be surrounded by those whose needs and desires he is able to satisfy. What they love in him is not himself, or his moral qualities, but his possessions. The great nobleman, therefore, may be considered a 'King of concupiscence' (620), surrounded by a small coterie of men and women who are the slaves of concupiscence. God, on the other hand, is the 'King of charity', surrounded by men and women who are full of loving-kindness. Thus, to be a great nobleman is to be in a situation almost antithetical to God's. At this point Pascal extends the analogy further than he had risked in the first *Discourse*, so that the realm of concupiscence now also includes all the actual kings of the earth. What makes for the strength of these earthly rulers is their 'possession of the things which men in their cupidity desire'.

Addressing now not merely the young future Duc de Luynes et de Chevreuse but, through him, all wielders of secular power, Pascal urges him to 'act as a true King of concupiscence': the fact that it is no virtue in the duke or king himself which gives him power over others should lead him to the humble realization that he has no right to tyrannize them. So much for the secular standpoint; but, from a religious one, not even this advice will be enough to gain salvation. By following all such precepts, the most a great nobleman or king can hope for is 'to damn himself, but at least to go to his damnation like a gentleman'. Then, at least, he will not have damned himself through avarice, lust, wrath or other deadly sins. But in order to be confident of salvation, he 'must scorn concupiscence and its realm, and aspire to that realm of charity whose subjects radiate nothing but charity and who desire only the good gifts of charity' (620–1).

The three *Discourses*, though little known (perhaps because they were not specifically written down by Pascal), are both large in their inspiration and bold in their execution. Though not actually composed by him, they are astonishingly like him and cannot be far removed from what he said. Above all, they foreshadow in so many respects the great work which Pascal had quietly been preparing for perhaps as long as three years before giving his instruction to the young Honoré-Charles d'Albert. This was the defence of Christianity to which those of Pascal's friends who were in the secret looked forward as if to some Messianic revelation. Even before the abrupt halt to the *Provincial Letters*, Pascal, health permitting, had been compiling notes and jottings for use in the ultimate writing of the Apologia. The *Discourses on the Worldly Condition of the Great* flow from the same state of mind which, over a much longer and more sustained period, was engrossed in the

Thoughts. There is a major difference between these works, in that the *Discourses* are concerned with a man of high estate whereas the *Thoughts*, infinitely more wide-ranging, are concerned with man. Yet in so many other respects they are identical in inspiration, and most of all in their accent on the strangeness, arbitrariness and deceptiveness of the human condition, its disquieting contradictions lulled into a semblance of rationality by the norms of social convention. In the world as a whole, concupiscence abounds: indeed, this is already clear from the *Discourses*. The distinction between two orders of greatness can be enlarged, along Pascal's own highly individual lines, very different from Rousseau's, into a wider conception of discontinuous orders. The stranger shipwrecked on some far-flung desert island is like Man born incomprehensibly on a far-flung planet. Both in the *Thoughts* as a whole and in the *Discourses*, especially the first, Pascal's leitmotiv is similar to that of the Epistle to the Hebrews: that we are 'strangers and pilgrims on the earth'.[41] It would be untrue to suggest that the *Discourses* have inspired the content of the *Thoughts*: rather the reverse. Pascal's instruction to Honoré-Charles d'Albert, important as he considered it to be to instruct kings and princes, stems from a far wider preoccupation with the plight of man, the danger of damnation threatening men and women of all sorts and conditions whether or not they have actually fallen into the deadly sins. The human condition of the rich and powerful, of which their worldly condition is but a shadowy likeness, is essentially similar to the whole human condition, though magnified to a critically dangerous extreme. And so, in the *Thoughts* as a whole, Pascal will refer contingently to the plight of the wealthy aristocratic man – his illusions, diversions, temptations and would-be satisfactions – whilst still pointing essentially to the eternal dimension in which perhaps the life of every man should be viewed.

7
Human Nature

In so far as any plan can be discerned in the *Thoughts*, it seems likely that the finished Apologia would have begun with an extended discussion not only of the human condition, in general and abstract terms, but of the personality of man. Pascal, as a contemporary of the *moralistes*, such as La Rochefoucauld, La Fontaine and La Bruyère, was well able to embark on this. Thus, the work of a seventeenth-century writer, misunderstood by Voltaire and others in the eighteenth century, was not destined to become widely appreciated and revered until the nineteenth century, whilst not perhaps until our own has it been fully understood in the light of modern psychological enquiry.

Addressing first and foremost the agnostic, but also the atheist and the deist, Pascal sees, most clearly of all things, that the primary requirement of his grand design is to stop the unbeliever in his tracks, shattering his complacency. In this he was noticeably successful with Voltaire, causing him to stop, ponder, and reassess his philosophical position. The sort of unbeliever he particularly has in mind is the rich, urbane and perhaps aristocratic gentleman, such as the Duc de Roannez, the Chevalier de Méré and Damien Mitton,[1] the type of man for whom all the affairs of the world go smoothly, happy in love, relaxed in conversation, proficient at games and sports, well versed in the social graces, including a polite smattering of literature, and perhaps also in the arts of war: the very *Mondain*, in other words, who eighty years later could have boasted in Voltaire's words: *The earthly Paradise is where I am*. Such a man would by no means necessarily be an atheist, but he would prefer the present to the Hereafter. For him, the present world is not only all he knows, but all he wishes to know, and all he feels that he really can know. Transcendent metaphysical speculation does not interest him; death is a remote abstraction, even though at all times – and, most of all, in warfare – it is a close reality. Nominally at least, he may count himself a deist: a believer in a distant, impersonal Creator almost if not totally divorced from the creation He has set in motion. In brief, his religion would be a natural one, not supernatural.

By talking to the freethinker about man (not the remote philosophical abstraction, but Everyman caught up in the web of day-to-day life with all its joys, pleasures and heartache), Pascal not only shatters his complacency, he also speaks to man about the one subject in which all men can be guaranteed to take a never-failing personal interest: themselves. In his great effort to prove the truth of the Christian religion, he clearly sees that the traditional theistic arguments – the argument of the First Cause, the related argument from design, the argument from Natural Law, the ontological argument, the Cartesian argument, and the various moral arguments which later were to be much favoured by Kant – are not enough. They are not enough because, at very most, all they will prove is the existence of a First Cause, or Prime Mover, or Supreme Artificer of the Universe: the initial Mind and Will imparting the first impetus to Creation, but not the incarnate God born in a manger, crucified on Calvary, risen, ascended and glorified. Deistic arguments based either on the harmonious functioning of the natural order, and the interrelationship of its innumerable parts, or else on the implantation of a sense of moral order within the human conscience may or may not incline human hearts towards a God; they cannot, however, bring out the peculiar distinctiveness of Christianity.

Pascal clearly realized that to defend Christianity in deistic terms would be tantamount to intellectual dishonesty. For the most that 'the spacious firmament on high' (the natural physical order) can reveal is the existence of a natural, not a supernatural, God. Against natural religion, centred on the natural order of the created universe and on the human instinct of natural justice, he defends revealed religion: the religion of a personal God approachable through prayer and the sacraments, in both individual and collective acts of worship (449*).

If, then, *God is love*[2] and if He is to be approached by man through prayer, penance and worship, the accent in the Apologia must fall squarely on man, the object of God's love. All Pascal's peculiar skills as a *moraliste* would doubtless have been brought to bear on those early chapters of the Apologia in which he would have spoken – and, in a fragmentary way, still does speak – to man of himself. These early chapters, writes Chateaubriand,[3] are 'above all remarkable for the profundity of their sadness, and for their strange immensity: you are suspended amidst Pascal's sentiments as it were in the infinite'.

Perhaps the clearest clue to his general method is given in *Thought* 130*:

> If he praises himself, I humble him.
> If he humbles himself, I praise him.
> And I keep on contradicting him
> Until he comprehends
> That he is a monster that is incomprehensible;

whilst, entering somewhat gingerly into the finer detail of Pascal's outline, *Thought* 6* is exceptionally revealing:

> First Part: Wretchedness of man without God.
> Second Part: Happiness of man with God.
>
> in other words
>
> First Part: Nature is corrupt, as is proved by nature itself.
> Second Part: There is a Redeemer, as is proved by the Scriptures.

From these two indications of an overall plan three major points emerge. In the first place, Pascal will take care to avoid all extremes in his presentation to man of the truth about his condition; and, similarly, he will take care to ensure that man avoids all extremes in the perception of the truth about himself. If the unbeliever forms too conceited an estimate of his own powers, he must be abased; if too lowly in his self-assessment, then he must be praised. At either end of the spectrum of human feeling, man must avoid complacency and he must avoid despair. He must not think that there is no hope for himself; but neither must he think that there is no need for hope.

In the second place, what *Thought* 6* reveals is that, whereas the *moralistic* study of man would have been designed to illustrate his wretchedness, the Apologia as a whole would have moved forward – as it were, from darkness into light – with a revelation of man's potential ultimate felicity. And the way in which it would have moved forward is the very path from nature into revelation: not, however, the nature of the natural physical order, with all the glories of its 'spacious firmament on high', but instead the nature of human nature, with all the miseries of its wretchedness, unhappiness and vanity. Moving, in the second part of the

Apologia, from a revelation about the human order to a revelation about the divine, Pascal would have sought to prove from the Bible, the revealed Word of God, that in Jesus Christ God became man in order to redeem man. Thus, in the deepest sense, both halves of the book would have constituted a revelation; for, to the self-satisfied libertine, the 'truth' (if such it is) about his actual human nature would in itself have been a revelation to him.

Moreover, from beginning to end of the Apologia, the message would have been about man: man first becoming aware of himself and his true nature; and man next being taught to think of himself as the object of God's Love and Redemption. But, just as Pascal knew that there would be no point in knocking on the door of the unbeliever in order to address him forthwith about arguments from design and First Causes, so too he realized that there would also be no point in his immediately addressing the unbeliever about the Bible. For who will accept the authority and relevance of the Bible unless he has first been made to see its importance and veracity? A crucial first stage on the way to accepting the 'truth' about God's redemption of man is, therefore, to face up to and to accept the truth about oneself.

This *truth about human nature itself* is the third major point to arise from Pascal's preliminary outlines of his intentions. To be human is to be a creature of paradox, so hopelessly incomprehensible that, quite contrary to the great maxim of Classical humanism, Γνῶθι σεαυτόν (Know thyself), a man cannot ever understand himself as man. And the recognition of his paradoxicalness is the tragic recognition of his monstrosity, that ἀναγνώρισις (anagnorisis) which, in Aristotle's view,[4] is of the very heart and essence of tragedy. What is more, as in some Racinian tragedy the protagonist (who is the unbeliever) will be pushed forward in an endless dialectic of thesis and antithesis towards an eventual recognition of the truth.

Thus, in the ultimate analysis, Pascal's goal is identical to the great adage inscribed at the temple at Delphi: *Know thyself*; except that a man can never learn to know himself by means of humanism alone. For, in Pascal's view, man is not self-sufficient but a creature of God, and cannot properly be understood without reference to a Hereafter. Two fundamental issues separate Pascal's outlook and, consequently, his defence of Christianity from that of humanism: the strictly limited importance he concedes to reason in religious and human matters, and his criticism of the human imagination.

Reason, he believes, is another thing of excess and paradox, towards which the only proper attitude is that of Μηδὲν ἄγαν (Nothing to excess), that other inscription written up at Delphi: the *juste milieu* which recognizes that truth lies midway between the twin extremes of too much reason (in this case, scientific deism) and none at all (i.e., superstition).[5] If, therefore, from the standpoints both of humanism and of science Pascal is surprisingly censorious of reason, he is still eminently Classical in his love of harmony and equilibrium – the *juste milieu*, as in a Molière comedy, between the excessive rationalizing of the so-called *raisonneur* and the opposite pitfall of having no reason at all.

In condemning the imagination Pascal seems even closer to Jansenism or Puritanism than to the Classical ideal. Imagination, in the *moralistic* opening of the Apologia, was to have been castigated as one of the three *puissances trompeuses* – the two other 'deceptive powers' being custom and vanity. Pascal points out (44*, 25*, 26*) how all men are misled by imagination into a false perception of truth. The imagination tends to distort and to magnify, lending to the object of contemplation or perception enhanced qualities which it does not actually possess. Thus, judges clad in scarlet appear – but only appear – to be wiser, more objective and more dispassionate than the rest of frail humanity (87*). It is because of their colourful robes that we willingly credit them with such superiority.

> There stands the greatest philosopher in the world on a plank that is wider than it need be; and there is a precipice beneath him. Although his reason may convince him that he is safe, his imagination will get the better of him . . . (44*).

> If [judges] administered true justice, and if doctors had the true art of healing, they would have no need of square caps. The majesty of those branches of learning would be sufficiently worthy of respect . . . (44*).

> We have only to see a barrister in cap and gown to form a favourable view of his competence . . . (44*).

Pascal is even scornful of the imaginative power of the artist (40*):

> The vanity of painting, which excites admiration through its resemblance to things the originals of which we do not admire!

Painting, however, is but one of the many forms of vanity treacherously inspired by the imagination: many more are inspired by custom.

In an age when tradition was far deeper-rooted than it is today, and when the habits and attitudes of the past were often treated with a respect bordering on reverence, Pascal is particularly bold in his attack on custom. He suggests that we too often believe a thing to be true merely because it has long been thought to be so (126*, 419*). The conventional wisdom is certainly not wise because it is conventional. As, for example, he himself had shown nine or ten years earlier in the matter of vacuums, the mere fact that a belief had been held since Classical antiquity is no evidence of its truth. Moreover, what is a custom at one time may not be so at another; but, more important still (a notion he has derived from Montaigne),[6] what is a custom in one place will not in another place be a custom at all: 'truth on this side of the Pyrenees, error on the other' (60*); so much for man's sense of natural justice!

Pascal's upbringing in a legal household[7] must also have made him intensely aware of another aspect of the prevalence of custom, the foundation unit of the French social structure in his time being the manorial court. Until August 1789 the manor, with its heriots, reliefs, escheats, merchets, chevages, and authority both criminal and civil, regulated the lives of the freeholders and copyholders who were its inhabitants. And in every manor, across the length and breadth of the kingdom of France, civil matters affecting the manorial tenants were regulated by *custom*: the particular custom of that lordship, so sacrosanct and invariable that it was written down into the court rolls. Thus, from time immemorial it may have been the custom of a manor to hold a certain parcel of land partly by virtue of a fixed payment in kind, whilst the descent of that parcel of land on the death of a sonless tenant might be to the daughters, one daughter, the eldest or the youngest brother as custom laid down. Such provisions no established power would (foreseeably) abolish or modify, yet in an adjacent seigniory the custom could well have been quite otherwise. What, therefore, Pascal is seeking to emphasize – in a climate of dutiful observance of hallowed custom, but custom varying even from village to village – is the relativity both of social custom and of 'natural' law.

Foremost amongst the deceptive powers, however, is imagination, which is even stronger than custom because custom itself feeds on it. It is ironical that Pascal should have written of the artistic imagination in such hostile terms when he himself employed so lively an artistic imagination to assert the fundamentally poetic, rather than scientific, truth of Christianity. In order to convince his readers of this truth, the man who appears so hostile to the artistic imagination in fact resorts to imagination in a high degree.

Man, therefore, in the process of self-knowledge becomes aware of himself as paradoxical: 'neither angel nor beast' (678*), a creature of 'grandeur' and 'wretchedness' (117*), deluded by vanity, custom and imagination, the 'glory and refuse of the universe' (131*). His 'grandeur' lies in the angel-like rational faculty which, for example, can achieve all the great discoveries of science and can actually (says Pascal) encompass the universe. One of the most justly celebrated of all the *Thoughts* is that of the thinking reed (200*):

> Man is merely a reed, the weakest thing in nature, but he is a thinking reed. There is no need for the whole universe to take up arms in order to crush him: a trace of vapour or a drop of water is enough to kill him. But even if the universe were to crush him, man would still be nobler than his killer, since he knows that he is dying whereas the universe knows nothing of its advantage over him . . .

Here, the accent is not only on the dignity of human thought as a means of rationally encompassing the universe but also on that tragic ἀναγνώρισις, or recognition, of the physical magnitude towering over the puny frailty of physical man. 'Man's grandeur', says Pascal (114*), 'is great in that he knows of his own wretchedness'. To this picture of human wretchedness, which to Voltaire appeared so devastating, not only do the *deceptive powers* contribute but also man's endless capacity for *divertissement*.

The thought on *divertissement* (136*), a word which combines within itself the twin meanings of 'diversion' and 'distraction', is one of the ten or twelve *Thoughts* written in a single burst of fluent inspiration; evidence enough that it is one of the more crucial thoughts in the whole of the projected Apologia. To what extent, if any, it universalizes some personal neurosis of Pascal is a moot point. He sees all men as the victims of *divertissement*, the tendency or temptation endlessly to seek distraction in the pleasures of the world. These pleasures, traditionally thought of in Christianity as a snare and a delusion, are condemned in the Litany of the Anglican Book of Common Prayer as 'the deceits of the world, the flesh, and the devil' and by Pascal and St Paul as 'concupiscence'. Pascal, however, goes far beyond the traditional condemnation of worldly pleasure to include within his *divertissement* any form of activity which is secular rather than God-orientated.

Why, he asks, do men go hunting, or courting, or to war, if not by means of these distractions to divert their mental attention from that facing up to the truth about oneself, or ἀναγνώρισις, from which alone (it seems to him) a genuine religious inspiration can come?

> When I have sometimes begun to ponder the various anxieties of mankind, the perils and the troubles which men face at Court, or in warfare which gives rise to so many quarrels, passions and bold and often wicked enterprises etc, I have often said that the sole reason for man's unhappiness is that he does not know how to sit quietly in his own room. If he knew how to enjoy staying at home, a man with enough to live on would never go off to sea or besiege some fortress; he would never spend so much money on an army commission if he could bear living in town all his life, and he only seeks the company and diversions of gambling because staying at home gives him no pleasure . . .
>
> This is why gambling and feminine society, war and high office are so eagerly sought after. It is not that they really bring happiness, or that anyone imagines that true bliss comes from hare-coursing or possessing the money that can be be won at gambling: no one would want these as gifts. What people want is not the soft, easy life that allows us to think of our unhappy condition, nor the dangers of warfare, nor the burdens of office, but rather the turmoil which takes our minds off that condition and distracts us. This is why we prefer the hunt to the kill.
>
> And this is why men are so fond of hustle and bustle. This is why prison is such horrible torture, this is why the pleasures of solitude are impossible to understand. And this indeed is the main joy of being a king, that people are constantly trying to distract him and provide him with every kind of pleasure. A king is surrounded by people whose only thought is to divert him and prevent him from thinking about himself. For, king though he is, he has only to think about himself to be unhappy . . . (136*).

Thus, at the heart of man considered simply as terrestrial man, without any reference to a Hereafter, lies a vast inner desolation, only made tolerable in earthly terms by a slavish devotion to the world. Worldliness, therefore, is no positive life-affirming delusion (if such an apparent contradiction in terms is even conceivable!) but a paltry and inadequate substitute for true inward peace. 'Thou hast made us for Thyself, O Lord', writes St Augustine in the opening page of his *Confessions*, 'and our hearts are restless till they rest in Thee': this maxim could well serve as the epigraph to Pascal's vast apologetic design. It is a maxim which could equally well be applied to Racine's Jansenist-inspired tragedy *Athaliah*, where the Baal-worshipping Queen of Judah is pursued in the innermost depths of her consciousness by the proto-Christian God and led by Him into the Holy Place of the Jewish Temple, and exclaims on her first appearance in the play how much she longs for 'That peace which I seek and which always eludes me'. In such mental restlessness and disarray a recognition of true religious values may, and to some extent does, dawn on Racine's unbeliever. Pascal, on the other hand, shows how such restlessness – far from being the painful threshold to a new outlook upon the world – can be flouted, ignored and (in strictly worldly terms) 'overcome' by recourse to such worldly delights as cards, dice, warfare, hare-coursing and women.

A further essential light is thrown on the presumed layout of the Apologia by *Thought* 12*:

> Men scorn religion. They hate it and are afraid it may be true.
> The cure for this is first to show that religion is in no way contrary to reason, but is worthy of reverence and respect.
> Next it must be made attractive; good men must be made to wish that it were true and then be shown that it is true.
> Worthy of reverence, because it has understood human nature.
> Attractive, because it promises true good.

In this view of the plan it must be shown at an early, if not the earliest, stage that 'religion is in no way contrary to reason', in other words, that it is not superstition. After which its attractiveness or desirability must be established and then, but only then, its dogmatic truth. It is, of course, already clear from *Thought* 6* that Pascal would have sought to prove the truth of the Redeemer by reference to the Scriptures; the distinctive feature of this additional statement is that he aspires to make the unbeliever yearn for a religion which if it were true would be desirable, before then proceeding to demonstrate its truth by Scriptural quotation. But how can this religion, desirable if it were true, be made to seem so attractive that the unbeliever craves for it with all his heart?

Clearly, the only means of proving the attractiveness and desirability of the Christian religion is to expose the utter inadequacy of alternative explanations of the meaning of human life. Not only does Pascal condemn the mindless hedonism of the man who is keen to have a good time, he also attacks the secular philosophical systems of stoicism, Pyrrhonism and epicureanism, the first two of which he had discussed with Isaac Le Maître de Saci some two or three years earlier in the *Conversation . . . Regarding Epictetus and Montaigne*. The mindless hedonism is futile because of the restlessness it creates but cannot dispel. As for stoicism and Pyrrhonism, Pascal had already dealt quite fully with these, in a fairly connected way, in the *Conversation*. Scepticism in particular incurs his fiercest rebuke, precisely because it is so destructive of the possibility of absolute and objective values. The first of the objectives of the Apologia is, after all, to demolish man's overweening confidence in himself. Pascal, as haunted by Montaigne's 'Apologia of Raimond Sebond' as Voltaire (and, later, Nietzsche) were to be by the *Thoughts*, does not subscribe to Montaigne's – or Charron's – belief that certainty is nowhere attainable in human life. For, if all were relative and nothing absolutely sure, what certainty could one then have of the existence of God? Indeed, would it be possible for God to exist at all?

Stoicism, at the opposite extreme of the gamut of secular philosophies, suggests as unduly high an estimate of man's powers as scepticism's estimate of his power to attain truth had been unduly low. In the writings of Stoic thinkers such as Epictetus and Marcus Aurelius man appears stronger, more resolute, braver in the face of adversity, than he in fact is. But the notion that man is the measure of all things only finds favour with Pascal to the extent that Christianity is intended specifically for man and is therefore expressed in language proportionate to man's understanding. In his view, man is a more subtle and complex entity than is allowed for in the ideal of Stoic humanism.

On what grounds, however, can it be said that religion succeeds where the secular philosophies fail? Pascal suggests, in the first place, that the paradoxical nature of man can only be explained by, and make its fullest response to, religion. The secular philosophies appeal to, and over-emphasize, one aspect of human nature to the detriment, or the exclusion, of another which is equally essential:

epicureanism appeals to the beast in man; stoicism to the angelic; scepticism places a huge question mark over the attainment of any certainty. It is, of course, no final and inescapably convincing proof of the authority of religion to say that, because religion is as full of paradoxes as is the nature of man, then religion must be true. All that can be said in its favour on this score is that, appealing to man, at least it does not over-simplify its appeal by failing to take account of the many-facetedness of human nature. Proving at most a negative, religion in its capacity for paradox is at least not manifestly wide of the mark. There is a suggestive and wholesome parallelism between its content and what is directly intuitable by man concerning the nature of his own experience. Compared with this, the content of the secular philosophies appears intolerably sterile.

There is, moreover, a second respect in which religion (according to Pascal) has a distinct advantage over the secular philosophies in its appeal to the human mind and heart. Although it is undoubtedly true that man cannot know everything – or perhaps even much – with certainty, even scepticism must admit that one thing at least is knowable with complete assurance: the fact of death. Thus, once again, Pascal adopts his by now familiar tactic of the Μηδὲν ἄγαν, the *juste milieu* between opposite extremes: man cannot know everything, but neither does he know nothing; scepticism is by no means entirely valid (in that at least one thing is knowable with certainty), but nor is it entirely invalid (since most of what men consider to be certainty belongs in fact to the realm of the 'deceptive powers' and is demonstrably *not* certain). But that death will come to each and every human being cannot be open to doubt. For, however possible it may be on logical grounds to dispute the fact of the inevitable death of every human being (Newman, in his *Essay in Aid of a Grammar of Assent*,[8] discusses this category of doubt), nevertheless, at a much deeper level of human cognition than that of formal logic – that of 'the heart [which] has its reasons of which reason knows nothing' (423*) – we all know that we shall die.

Death, the one certain event in the existence of every man, is at the same time the event most uncertain in its consequences. Scepticism, unacceptable in respect of the inevitability of death, is practically insuperable in respect of the existence and nature of an afterlife. 'No choice', Dag Hammarskjöld has written,[9] 'is uninfluenced by the way in which the personality regards its destiny, and the body its death. In the last analysis, it is our conception of death which decides our answers to all the questions that life puts to us'. It is possible, in other words, for the self to consider its own physical extinction in more than one way: the moment of death may or may not represent the total dissolution of an individual human identity; but whatever the 'conception of death' which finally 'decides our answers to all the questions that life puts to us', it will only have been arrived at by a transcendence of scepticism.

Before the willingness to overcome a natural openness of mind can ripen into positive commitment, Pascal must fully display the paradoxical nature of man. Having been made to 'comprehend that he is incomprehensible', man must now be brought to understand the ultimate paradox: that 'it is harder to conceive of man without this mystery [the mystery of the Christian revelation] than it is for man to conceive of this mystery itself' (131*). However paradoxical and mysterious Christianity may seem, mankind *without* Christianity would seem more paradoxical and mysterious still. Admirably, therefore, Pascal turns the tables on his freethinking or indifferent opponents, placing the onus of accounting for the greater and the lesser mystery upon them.

The greater mystery (in Pascal's view) of the human situation is that man is betwixt and between in the order of creation, a finite creature at the intersection of two infinites, and with features of both but the identity of neither.

> Let man then contemplate the whole of nature in her full and lofty majesty, and let him turn away his gaze from the lowly objects that surround him; let him behold the dazzling light set like an eternal lamp to illuminate the universe, let him see the earth as a mere speck compared to the vast orbit described by that star, and let him marvel that this vast orbit itself is no more than the tiniest point compared to the orbit described by the stars revolving in the firmament. But if our eyes stop there, let our imagination proceed further; it will tire of conceiving things before ever nature tires of producing them. The whole visible world is but an imperceptible dot in the great vastness of nature. No idea comes near it; in vain do we inflate our conceptions beyond imaginable space, we only bring forth atoms compared to the reality of things. It is an infinite sphere whose centre is everywhere and whose circumference is nowhere. In short, it is the greatest perceptible mark of God's omnipotence that our imagination should lose itself in that thought . . .

But, as evidence of another prodigy that is equally astounding, let him look for the tiniest things he knows of. Let a mite show him how in its tiny body there are incomparably smaller parts: legs with joints, veins in its legs, blood in its veins, humours in the blood, drops in the humours, and traces of vapour in the drops; let him divide these latter things further still until he has exhausted his powers of imagination, and let the last thing he comes down to now be the subject of our discourse. Maybe he will think that this is the ultimate of minuteness in nature.

I wish to show him a further abyss. I wish to depict to him not only the visible universe, but all the immensity of nature that can be conceived of within this miniature atom. Let him see that here there is an infinity of universes, each with its firmament, its planets, its earth, in the same proportions as in the visible world, and that on those earths there are animals, and even indeed mites, in which he will also find the same characteristics as in the earlier ones; and, finding the same characteristics again and again in these others without end or respite, he will be dumbfounded by such wonders, as astounding in their minuteness as the others are in their immensity. For who will not be filled with astonishment that a body which a moment ago was imperceptible in a universe that itself was imperceptible within the great vastness of *everything* should now be a colossus, a world, or rather everything, when compared to the nothingness which is beyond our reach? Anyone considering himself in this way will be terrified at himself; and, seeing how his mass, as given him by nature, supports him between these two abysses of infinity and nothingness, he will tremble at these marvels. I believe that, as his curiosity changes into amazement, he will be more inclined to contemplate them in silence than to investigate them with presumptuousness.

For what, after all, is man within nature? Nothingness in relation to infinity, yet everything in relation to nothingness; a middle point between nothing and everything, yet infinitely remote from an understanding of the extremes; the purpose and the principles of things are insuperably concealed from him in impenetrable secrecy.

Equally incapable of seeing the nothingness from which he has been drawn and the infinity in which he is engulfed.

What else, then, can he do but perceive some semblance of the middle of things without ever having any hope whatsoever of

> knowing either their principles or their purpose? All things have emerged from nothingness and are being carried forward towards infinity. Who can follow these astonishing processes? The author of these wonders understands them. No one else can (199*).

Strangely poised between the macrocosm of astrophysics and the microcosm of molecular biology, man is a creature of disconcerting and unfathomable contrasts. Pascal's insights are prophetic. Twentieth-century astronomy suggests that the planet on which we live is only one of a million planets within our galaxy alone, many perhaps inhabited by intelligent species; within the Milky Way galaxy are no less than 180,000 million suns, many or perhaps all of them having their own solar systems of revolving planets and satellites; whilst beyond our galaxy lie numerous other galaxies, the spiral nebulæ at least 100,000,000 in number, some larger than the Milky Way. As for the distances between our galaxy and the others, it may very well be that these, in an expanding universe, are constantly increasing; in time, given the unvarying velocity of light, even those galaxies which now can be seen will have vanished into the unobservable; and the Milky Way will appear to be alone in space, though it will not be alone; nor will it be at the centre of space, though it will seem to be so. This would mean, as Pascal surmised with an uncannily probable accuracy, that the universe has *its centre everywhere, and its circumference nowhere*: the repudiation of man's privileged, geocentric position within the cosmos can go no further.

Hardly less prophetic is his view of the composition of the tiniest particles of matter. Though he himself does not use the terms 'molecules', 'atoms', 'atomic nuclei', 'protons', 'neutrons' and 'electrons', he is indistinctly aware of that strange phenomenon of the physical universe, worlds within worlds: the universe viewed as a miniature solar system with negative electrons (six in the case of carbon, nine in the case of fluorine) orbiting around a positive nucleus. Within this picture of worlds within worlds there is, admittedly, an element of poetic exaggeration: but it is substantially the same element of exaggeration as is to be found, half a century or so later, in Leibniz's *Monadology* and related writings where matter is conceived of as being 'a *multitude* of different substances, like (as it were) a flock of sheep'[10] – although not infinitely divisible since Leibniz's ultimate unit is the monad.

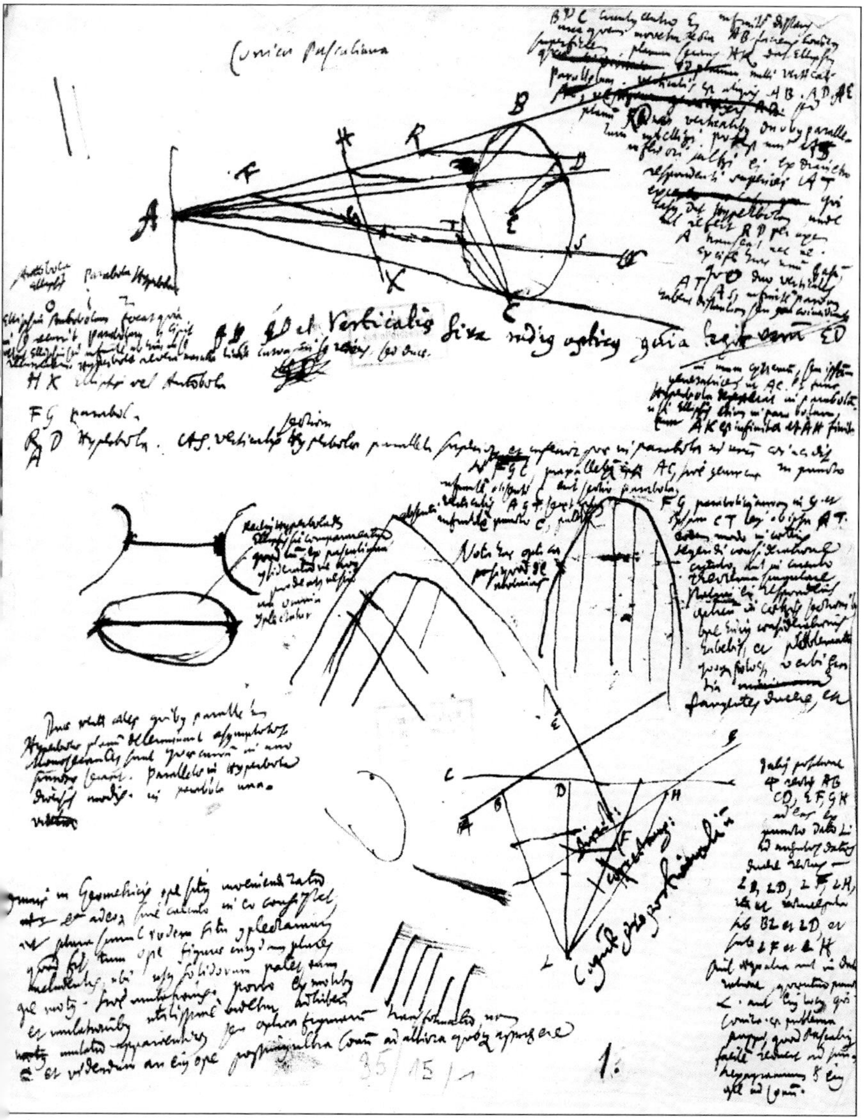

DRAWING OF PASCAL'S THEOREM OF THE MYSTIC HEXAGON
as recorded by Tschirnhaus for Leibniz in 1676;
the small (and predominant) handwriting is that of Leibniz.
(by courtesy of the Niedersächsische Landesbibliothek, Hanover)

2

PASCAL'S CALCULATING-MACHINE
formerly in the possession of Queen Marie-Louise of Poland.
Unlike Pascal's other mechanical calculators, this is capable
of all four arithmetical processes.
(by courtesy of the Staatlicher Mathematisch-Physikalischer Salon, Dresden)

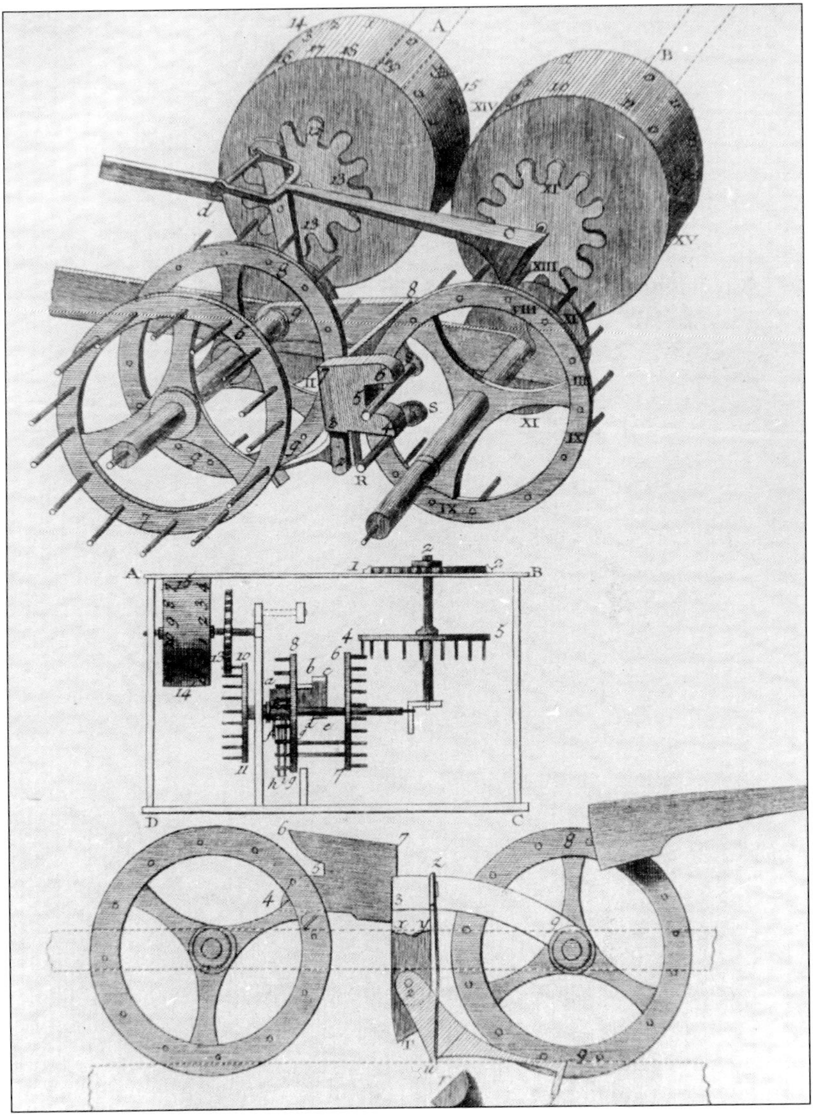

3 THE MECHANISM OF PASCAL'S 'ARITHMETICAL MACHINE'
from *Planches*, Vol. 1, of the *Encyclopédie* (3rd edn), 1779.
This took three years to perfect.

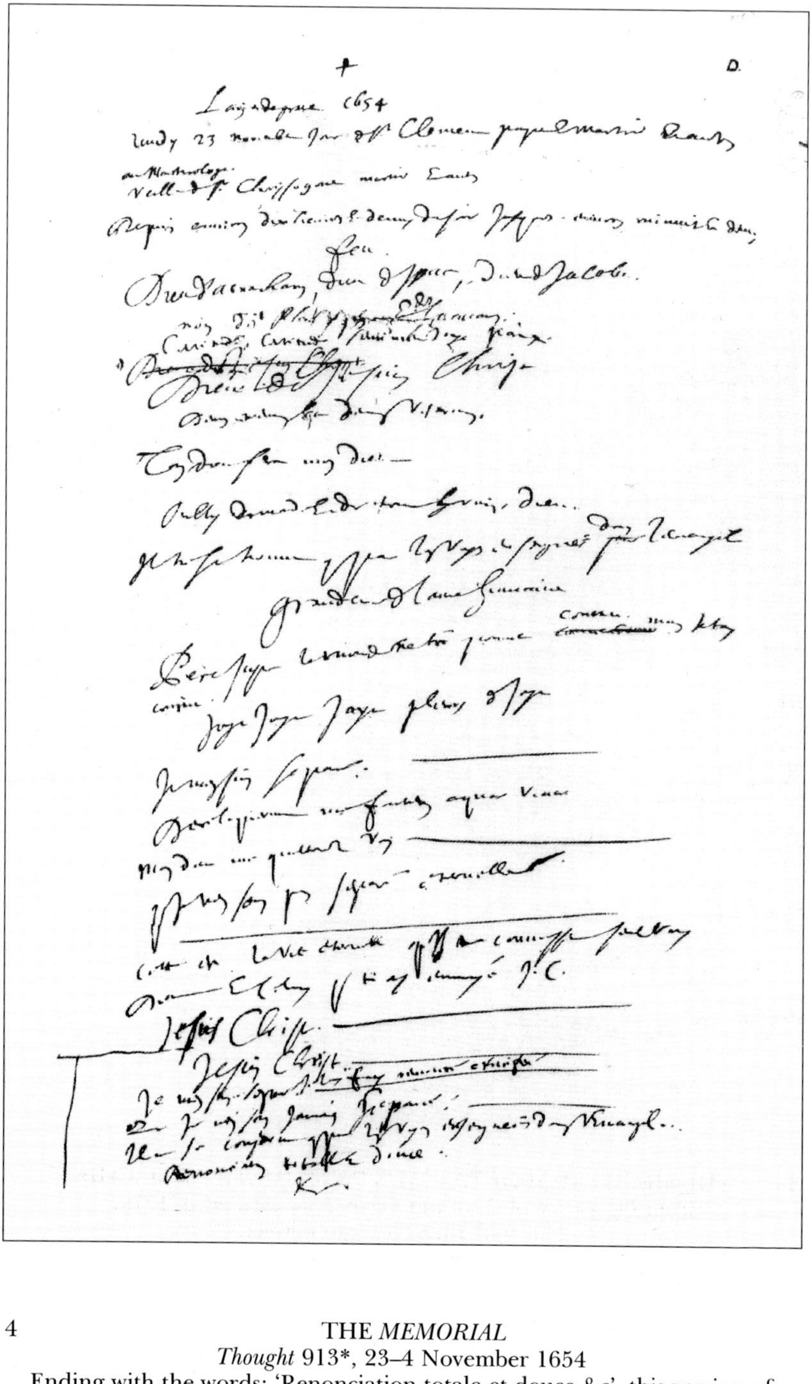

4 THE *MEMORIAL*

Thought 913*, 23–4 November 1654

Ending with the words: 'Renonciation totale et douce &c', this version of *Thought* 913* lacks the last three lines. It is in Pascal's handwriting, few specimens of which survive.

(by courtesy of the British Library, London)

5 MARGUERITE PÉRIER
(1646–1733)
Daughter of Pascal's sister Gilberte.
This painting was once reputed to be the work of Philippe de Champaigne.
Its Latin inscription reads:
TO CHRIST THE SAVIOUR
THIS PORTRAIT OF MARGUERITE PÉRIER, A TEN-YEAR-OLD GIRL WHO SUFFERED FOR THREE YEARS FROM A HIDEOUS AND INCURABLE FISTULA OF THE LEFT EYE BUT WHO ON 24 MARCH 1656 WAS INSTANTLY RESTORED TO HEALTH ON CONTACT WITH THE LIFE-GIVING THORN, IS THE VOTIVE OFFERING OF HER PARENTS IN COMMEMORATION OF SO GREAT A BLESSING.

Marguerite is said to have been cured by the Miracle of the Sacred Thorn, 24 March 1656. From a portrait in the Church of Linas, Essonne.

6 PASCAL'S WELL IN THE FARMYARD AT PORT-ROYAL DES CHAMPS
Its mechanism is said to have been designed by Pascal, possibly in May 1658.

7 PASCAL'S DEATH MASK

(by courtesy of the Principal and Fellows of Newnham College, Cambridge)

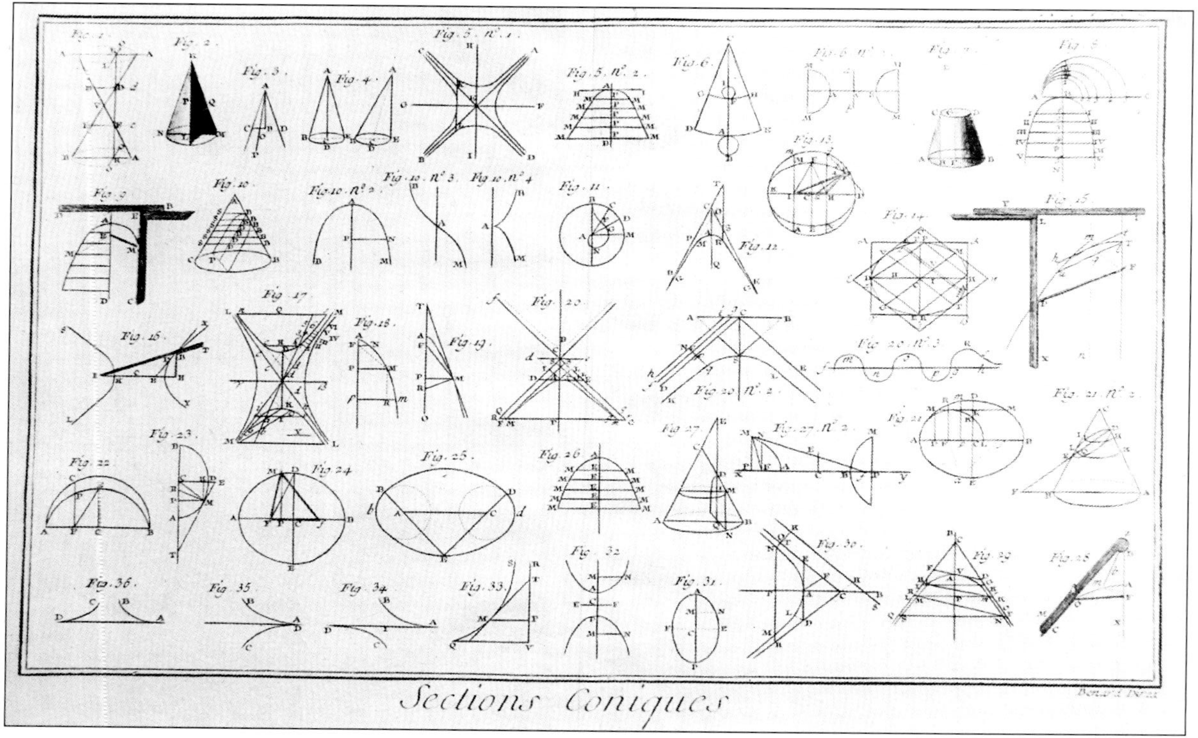

8 CONIC SECTIONS
from *Planches*, Vol. 1, of the *Encyclopédie* (3rd edn), 1779.
(copyright Donald Adamson)

By no stretch of the imagination could it have been *scientifically* thought, in 1658, that firmaments, stars, planets and animals existed *within* a mite. Electricity, not thoroughly investigated until the work of Franklin and Volta but already touched on by Guericke and Boyle, was a concept of which Pascal was unaware. The scientific research of J. J. Thomson and others has since shown that the atom and the molecule do not consist of particles of matter infinitely divisible into smaller and smaller worlds. Nevertheless, though the whole physical universe is ultimately resolvable into pulsations of electricity, Pascal's essential point remains undiminished: that man and the physical universe he inhabits are in no sense sufficient unto themselves but, though finite and material up to a point, hover mysteriously between two immaterial infinites.

Thus, placed between these extremes of the physical realm of extension, and placed also between the moral extremes of grandeur and wretchedness, man – a 'thinking reed' – is in one sense far greater than the universe yet in another sense far smaller than it; his great gifts of reason are liable to be led astray by the 'deceptive powers' (45*); and no amount of *divertissement*, that most deceptive of all occupations, can avert the fact of death even though it may blind him to it. It is at this juncture that the most crucial and controversial of the *Thoughts* enters into play: Pascal's concept of the Wager. Not having any final and positive means of determining the nature of the Hereafter, do we regulate our lives as *if* a Hereafter exists? or do we, on the other hand, regulate them as if there were nothing beyond the grave? And if we regulate our lives on the hypothesis of the existence of a Hereafter, what sort of an afterlife do we presuppose? Concerning a matter of such indubitable importance to us, if some kind of an afterlife exists, how can any certainty of it be attained? Above all, should we or should we not assume that a Heaven and a Hell, or some other future state, do exist?

The certainty to be arrived at, if any is attainable at all, is not of course the formal (and tautological) certainty of mathematical logic, but rather some form of practical operational certainty to be tried, tempered and renewed in the adventure of life. There is, as Pascal insists in the celebrated thought on the *esprit de géométrie* and the *esprit de finesse* (512*), a world of difference between *deductive* and *inductive* proof, between *a priori* and *a posteriori* reasoning, or as he himself expresses it:

> The reason why certain intuitive minds [*esprits fins*] are not mathematical [*géomètres*] is that they are quite unable to apply themselves to the principles of mathematics, but the reason why some mathematicians [*géomètres*] are not intuitive [*fins*] is that they cannot see what is before their eyes and, being used to the summary, self-evident principles of mathematics and being furthermore in the habit of drawing no conclusions until they have clearly seen and handled their principles, they lose their way in matters requiring intuition, in which principles cannot be handled thus. They [these matters] can hardly be seen; rather than being seen they are felt; and there is endless difficulty in imparting them to people who do not feel them for themselves . . . The thing must be seen all at once, at a glance, and not by virtue of progressive reasoning, at least up to a point.

Mathematics, at the most, could only prove the existence of a God Who was of the mind rather than the heart. Weaker minds might not seize the force of the arguments used in proof of His existence, just as the mind untrained in mathematics and physics may not apprehend the arguments in support of Quantum Theory, or Heisenberg's Indeterminacy Principle, or the First Law of Thermodynamics. But to minds capable of such understanding, at least it can be said that the First Law of Thermodynamics and all other laws of mathematics and physics are universally acceptable – and so will remain until (if ever this should occur) they are replaced by a more refined scientific model.

Even at the level of deductive, *a priori* proof there was, and still is, no general agreement as to the existence of a God. Realizing this long before the fact was made doubly obvious by relaxations in the climate of religious opinion which date only from the nineteenth and twentieth centuries, Pascal turned this philosophical difficulty to great advantage. In his concept of the *deus absconditus* (or Hidden God: the phrase is borrowed from Isaiah XLV 15),[11] he was again two centuries in advance of theological thought. *Verily Thou art a God That hidest Thyself*: hidden in a manger, in a carpenter's shop, in the scandal of the Cross, and in the tomb; concealed from man in the sorrows of the Suffering Servant.[12] It is this argument, that there is not too much light of the divine Revelation but neither is there too little, on which Pascal laid particular stress in the lecture notes intended for a talk delivered at Port-Royal des Champs towards the end of May 1658: this progress report (*A.P.R.: At Port-Royal*) was two or three hours long[13] and was said to have been so eloquent and wide-ranging as to leave all his listeners 'spellbound with admiration'.[14] 'It was not therefore right', the notes conclude,

> that [Jesus] should appear in a manner that was manifestly divine and absolutely capable of convincing all men, but neither was it right that His coming should be so hidden that He could not be recognized by those who sincerely sought Him. To these people He wished to make Himself perfectly recognizable; and, thus wishing to appear openly to those who seek Him with all their hearts and yet remain hidden from those who shun Him with all their hearts, He has tempered our knowledge of Him by giving signs of Himself which can be seen by those who seek Him but which cannot be seen by those who seek Him not.
>
> There is enough light for those who desire only to see, and enough darkness for those who are of a contrary outlook (149*).

Yet again, therefore, Pascal resorts to what is his most characteristic philosophical device in the *Thoughts*: the Μηδὲν ἄγαν, or *juste milieu* between extremes. Had the facts of the Christian message been made blindingly self-evident to all, this could only have amounted to the revelation of natural religion; whilst no disclosure of them would have been a plain encouragement of atheism. There must be enough light, but not too much nor too little: for to shed too much would be to incline towards reason, whilst to shed too little would be to incline towards superstition. The kind of conversion or commitment sought by Pascal is not such as can be elicited by the blinding compulsion of historical or mathematical proof.

More than two centuries after the writing of the *Thoughts*, Newman in his *Essay in Aid of a Grammar of Assent* echoed Pascal's outlook – though voicing it more clearly – when he remarked how difficult, and often indeed impossible, it is to draw out the grounds on which the human mind yields that conviction, or assent, which in his view admitted of no degrees and was either entire or did not exist at all. Christianity, he argued,[15] addresses the mind

> both through the intellect and through the imagination; creating a certitude of its truth by arguments too various for enumeration, too personal and deep for words, too powerful and concurrent for reversal. Nor need reason come first and faith second (though this is the logical order), but one and the same teaching is in different

> aspects both object and proof, and elicits one complex act both of inference and of assent.

One of Pascal's merits is to have been the first of all Christian apologists to realize, as Newman later did, that 'powerful and concurrent' reasons contribute to our acceptance of the truth of Christian teaching. Yet he differs from his successor precisely in the emphasis given to the thought on the Wager – arguing it more cogently and at greater length than any other *Thought,* and doubtless regarding it as more 'decisive' than 'concurrent' (though having characteristics of both). 'We know the truth not only through our reason but also through our hearts', writes Pascal (110*), anticipating by two hundred years Newman's spiritual motto *Cor ad cor loquitur* (*Heart speaks to heart*).[16] And it is to the heart, and its 'instinct and feeling' (110*), that Pascal essentially appeals when by careful degrees he brings his reader to the point of sharp decision at which it is no longer a question of the middle way, the harmonious compromise between opposing principles: now, with the Wager, comes (as sooner or later death will come) the stark choice between belief and disbelief in an afterlife.

8

The Wager

To a man for whom life was so obviously a heroic adventure (though he lacked any physical capacity for heroism), any tactics for the defence of Christianity must ultimately be founded upon risk-taking. 'St Augustine', writes Pascal (577*), 'saw[1] that we take risks at sea, in battle, etc. – but he did not see the rule of probability which proves that we ought to do so': he lacked, in other words, an understanding of the probability calculus on the basis of which human beings rationally act. Pascal likewise invokes the risks incurred by anyone walking along a plank (44*). Where, he believes, his superiority over St Augustine lies is in his work on probability theory jointly undertaken with Fermat in the summer of 1654, and by which he was still engrossed in September 1656 and in March 1657. Armed with this mastery of probability theory, he engages his freethinking or indifferent opponent in a dialectic so relentless that it is actually couched in dramatic form: the only occasion on which he proceeds by question and answer in the whole of the *Thoughts*.

'God either is or He is not': Pascal opens the debate challengingly, if with rather deceptive simplicity. On which alternative, then, will the agnostic gamble (if indeed there is any need to gamble)? Reason, Pascal adds, will be of no assistance to him in deciding how to place his bet.

To which, predictably enough, the agnostic replies that the only proper course is not to wager at all.

Pascal's whole purpose, however, is to goad the agnostic out of his slothful indifference. He insists: ' . . . You must gamble. This is not a matter of free choice. You have embarked . . . '

Thus, the individual human life, rather like the physical universe suspended between nothingness and infinity, is presented as a sea-voyage from an uncertain origin towards an uncertain goal. Yet the fact that man is uncertain of the goal does not mean that he should refrain from suppositions concerning it. On the contrary, Pascal now goes on to outline six hypothetical gambling situations, all of them intended as possible models of the gambling risk a man either must, or need not, take in life.

In the first situation, the probabilities of gain and loss are equal (God either is or He is not) and the stake is supposedly nil (no sacrifice of one's life being involved), so that the hypothetical loss is nil whilst the gain is supposed to be *everything* (which may or may not, and probably does not, amount to *infinity*). In the second situation, which has two slightly varying formulations, the probabilities of gain and loss are again equal, the stake is finite (one's own life), and the gain is also finite (two lives or three). In the third situation, which again has the same two slightly varying formulations, the stake is finite (one's own life), the gain is also finite (two lives or three), and the probability of gain is one out of 'an infinite number of chances'. So far, it seems, Pascal has only considered situations in which the possible gain is finite.

In the fourth situation, on the other hand, the hypothetical gain is infinite, but the stake is finite (one's own life), and the probability of gain is 'one chance of success against a finite number of chances of loss'. In the fifth, as in the second situation, the probabilities of gain and loss are equal, and the stake is also finite (one's own life); but, as distinct from the second situation, the hypothetical gain is again supposed to be infinite. Finally, in the sixth situation, the prospect of gain is no longer a probability but a certainty, whilst the stake is nil and the gain infinity.

Expressed algebraically: if p be the probability, n the number of chances, s the stake, q the quantity risked, x the gain, and z 'everything', then

in the first model, $p = \frac{1}{2}$, $s = 0$, where $x = z$ (OR ∞?)
in the second, $p = \frac{1}{2}$, $s = q$, where $x = 2q$ or $3q$
in the third, $p = \frac{1}{\infty}$, $s = q$, where $x = 2q$ or $3q$
in the fourth, $p = \frac{1}{n}$, $s = q$, where $x = \infty$
in the fifth, $p = \frac{1}{2}$, $s = q$, where $x = \infty$
in the sixth, $p \approx 1$, $s = 0$, where $x = \infty$.

But a wager in which $p \approx 1, s = 0$ is, of course, no wager at all, nor in fact does the situation $p = \frac{1}{2}, s = 0$, where $x = z$ (OR ∞?) constitute a wager as properly understood. For, according to the normal definition of a wager, a stake must be put down and odds must be laid (the ratio between the amounts staked often, but by no means always, being 1:1). To stake nothing is not to participate in a wager at all. Likewise, when the prospect of success is a certainty (as in the sixth model), no risk is taken. Moreover, a wager is entered into by two or more parties each of whom stakes a finite quantity (usually of money) against the others' on the outcome of a doubtful event. Against whom is the agnostic being made to wager? Not against Pascal, for the hypothetical gain is infinite in at least three out of the six models, 'everything' in perhaps one, and a quantity of multiple lives in the other two: at all times it is a gain which it is impossible for Pascal to pay. The agnostic, in fact, is being made to wager against God – a God Who may or may not exist!

Out of the six models of the Wager, only one (the second) comes close to the wagering situation as commonly understood. In the fourth and fifth models the stake is finite, as is the probability; the prospective gain is, however, infinite: this is a unique feature amongst wagering situations. The probability in the third model is a fraction whose divisor is ∞, i.e., nil. There is no stake in the first and sixth models. All six are uncommon wagering situations in so far as the agnostic is being compelled to wager: he does not wager of his own free will.

In what are the three nearest approaches to the conventional model of a wager Pascal first takes the situation $p = \frac{1}{2}$, $s = q$, where $x = 2q$ or $3q$ in which the prize to be won is merely finite, but where there is an even chance of winning it: this, though the most mathematically acceptable of the various models, is remote from his real purpose which is to persuade man to stake the finite for the infinite. The first ($p = \frac{1}{n}$, $s = q$, where $\frac{1}{n} = \infty$) of the wagering situations in which infinity is the hypothetical gain would clearly cause any wagerer to think twice if his primary concern was to minimize possible loss; but – having, at any rate to his own satisfaction, demonstrated that the risk in the fourth model is an acceptable risk to take – Pascal proceeds to argue that in the fifth model ($p = \frac{1}{2}$, $s = q$, where $x = \infty$) the wagerer's decision is an easier one still since he has an even chance of winning. All these models, but particularly the fourth, are of course disturbing to the wagerer because of his fear that q (the quantity risked) may far outweigh the possible gain.

Indeed, at the very outset of the debate Pascal had recognized the force of this objection by claiming ($p = \frac{1}{2}$, $s = 0$, where $x = z$) that the wagerer was actually wagering nothing. In reply to the agnostic's first protest, that he did not wish to wager at all, Pascal had insisted that there was no alternative for him but to do so: 'you have embarked'; moreover, 'nothing' was being risked whilst there was an even chance that 'everything' (z) would be gained. When, in answer to this, the cornered agnostic had complained that the stake might still be unacceptably high, Pascal had abruptly changed tack. He had now argued that where there was an even chance of success and where the wagerer was compelled to wager anyhow, it was prudent to stake one finite life in the hope of gaining two or three ($p = \frac{1}{2}$, $s = q$, where $x = 2q$ or $3q$):

> Let us see: since there is an equal chance of gain and loss, if you stood to win only two lives for one you could still wager. But supposing you stood to win three? You would have to play (since you must necessarily play) and it would be unwise of you, once you are in a situation where you have to play, not to risk your life in order to win three lives at a game where there is an equal chance of winning and losing.

However, though the agnostic had made no reply (indeed, throughout the greater part of Pascal's discourse he does not manage to get a word in edgeways), the second model had not satisfied him. Being more concerned to minimize a possible loss than to maximize any expected gain, he was disinclined to renounce all worldly joys and delights for – perhaps the blank nothingness of death. He could not agree that it was worth while to hazard his present existence, a sweet and enjoyable thing to him, for some intangible and chimerical afterlife. At the same time, the notion of risking one life in order to gain two or three would not have appeared so foreign or quixotic to him as it might appear to the reader of today. Estates of inheritance were a common legal phenomenon in the seventeenth and eighteenth centuries, and were frequently prolonged by adding to them one or more lives.

Thus Pascal brings his agnostic face to face with the core of his argument, according to which the potential gain in the Wager would be 'an eternity of life and happiness'. First, however, he proposes a model of the wagering situation ($p = \frac{1}{\infty}$, $s = q$, where $x = 2q$ or $3q$) which may seem somewhat extravagant. According to this formulation of the odds, where the stake is again the wagerer's own life, only one chance of attaining a gain of two or three lives is allowed against $\infty - 1$ chances of failing to do so. $\infty - 1 = \infty$, and there are therefore an infinity of chances against the wagerer. Nevertheless, Pascal urges, the daunting immensity of the risk would be worth taking even for a prospective gain of two or three lives. 'You would still be right to wager one', he assures the agnostic, 'in order to win two'. He would be wrong not to wager, Pascal assures him, if three were the prospective gain. And the reason why, according to Pascal, he would be wrong not to wager is that he is 'obliged to play'.

The agnostic remains silent but is unconvinced.

This third model of the wagering situation, however unconvincing it may be not only to the mathematical layman but also to the specialist, seems in fact intended to serve as a mere foil to the fourth, where the stake is the same as in the third though otherwise the terms are reversed: now the prospective gain is ∞ rather than the finite $2q$ or $3q$, whilst the probability if no longer $\frac{1}{\infty}$ but the finite $\frac{1}{n}$. For if the odds are acceptable in the third model, how much the more so must they be in the fourth! This fourth model together with the fifth are the very crux of the argument of the Wager. Narrowing the odds, Pascal first makes the ratio of the probability $1{:}n$, then eventually 1:1; and all within a few lines.

> If there were an infinity of infinitely happy life to be won: but here there is an infinity of infinitely happy life to be won, one chance of winning against a finite number of chances of losing, and what you are staking is finite. That leaves no choice; wherever there is infinity, and where there are not infinite chances of losing against the one chance of winning, there is no room for hesitation, you must give everything. And thus, since you are obliged to play, you must be renouncing reason if you keep your life rather than risk it for an infinite gain, just as likely to occur as a loss amounting to nothing.

The whole of the next paragraph amplifies the bold claim made by Pascal in this last sentence.

Here the logical objection which he is anxious to repudiate is that the uncertainty of the gain invalidates the Wager. Every gambler, he points out,

> takes a certain risk for an uncertain gain, and yet is taking a certain finite risk for an uncertain finite gain without offending against reason.

Mathematically, he argues, this case is indistinguishable from the one in which the gambler plays for an uncertain infinite gain, because the ratio of the *uncertainty* of winning – whether it be a finite or an infinite prize – to the *certainty* of what is risked is as the ratio of the chances of winning to those of losing. Consequently, if it can be assumed that there are as many chances on the one side as on the other ($p = \frac{1}{2}$, $s = q$, where $x = \infty$), the agnostic will be playing for even odds. For once Pascal has found an argument that convinces even his agnostic, who – to tell the truth – is much more readily persuaded than many of the subsequent critics of the *Thoughts* have been.[2]

In the first place, the Pascalian Wager is open to the obvious objection that it is not constructed on the normal model of a wager at all. Of course, it can be replied, this is not a normal wagering situation: the reluctant gambler is placed in exceptional circumstances; the prize which stands to be won is exceptional in that, if it exists at all, it is infinite; the wagerer is actually wagering on the possible existence of the prize; and, what is more, he is being compelled to make a bet. But need he bet at all? The fact that the unbeliever is being compelled to play does not necessarily imply that he is being compelled to stake. And what is meant by the nine times repeated assertion that the unbeliever is being *compelled to play*?

This is the kernel of Pascal's argument, and one which invalidates the usual concept of a wagering situation. For in any normal wager the risks are undertaken of the gambler's own free will. If the unbeliever is being compelled to wager, in what sense can he gamble except to place his bet on the existence of an afterlife? No one, within the simple terms of a bet, would stake his life on the possibility that there is not an afterlife. This, therefore, illustrates Pascal's unargued assumption that by pursuing the seductions of 'the world, the flesh, and the devil', one is betting that there is no life beyond the grave. But to consider the stake of the (enforced) Wager as being also its gain seems to defy every principle of the law of probability. Clearly, it is possible in ordinary circumstances to keep one's £1 rather than place it on a horse. Then, however, the £1 is no *gain*: it is only the certain absence of a possible loss. All conventional – and, many would add, meaningful – models of the wager are based on the assumption that that £1 is worth keeping, and the strong inducement to parting with it is that thereby one may gain many times more of the same thing. Throughout his argument of the wager Pascal is intent on persuading his unbeliever to maximize gain rather than to minimize loss. But supposing the unbeliever, relishing his present existence and seeing that in the argument of the Wager like is not being compared with like, is determined to minimize a loss he can understand rather than to maximize a gain he cannot visualize?

Pascal's Wager differs, therefore, from any conventional model of the gambling situation in that the stake is the whole life, whereas the £1 laid on a horse is but a tiny portion of all that goes to make up the conventional gambler's life: and the gambler (unless he was a compulsive gambler) would be less and less inclined to part with the £1 in proportion as that sum became a larger and larger part of the total of his life's assets! This prompts the reflection that perhaps the Pascalian Wager should also be conceived historically or diachronically, in which case there would be no need for the unbeliever to embark on it today when it could be delayed until tomorrow, thus involving a lesser sacrifice of the gambler's earthly life. Pascal does not, however, conceive of his Wager in diachronic or historical terms. He views it as a momentous challenge, of eternal import, to be faced immediately.

There is also, on the other hand, a second sense in which his Wager differs from any conventional gambling situation: a fact which brings us on from the second and third models of the Pascalian argument to the fourth and fifth. For these models, unlike their two immediate forerunners, presuppose that the stake itself is fundamentally without value. The symmetry of the argument now becomes obvious, as Pascal proceeds from a situation in which the stake has no mathematical value (though the life which is *not* staked has a real value to the unbeliever who calls heads or tails), to two situations in which the stake has both a mathematical and a real value, to two in which the stake has a mathematical value but an *unreal* one in that the prospective gain is of a different order of magnitude, to a final situation in which the value of the stake is *nil* both in mathematical and in existential terms. Hard as it may be to proceed from the first to the second and third models for any wagerer keen to minimize his potential loss, it is harder still for a wagerer bent upon maximizing his gains to proceed from the third model to the fourth. This is because, in mathematical terms, the prospective gain no longer bears any recognizable resemblance to the stake – and also because of the implicit assumption, in the fourth and fifth models, that that stake is fundamentally valueless. To Pascal it is valueless because of the ephemeral and delusory nature of all worldly delights. He believes that, in order to stand a chance of gaining the infinite, it is not enough for the unbeliever to make a hollow profession of faith: such lip-service is for the Jesuits alone. As he sees it, religious belief entails moral consequences such that the unbeliever's whole lifestyle would be revolutionized: the wagerer would come to feel scorn for worldliness and *concupiscence*. But the stake in the fourth and fifth models is by no means valueless to the unbeliever himself! Far from staking nothing, he would be staking all the happiness and satisfactions of a worldly life. And supposing there were nothing beyond the grave? . . . The lack of any recognizable relationship between the stake and the gain is, at the very least, an almost insuperable obstacle to those wagerers for whom the maximization of gain is the first priority; the value that the stake still has for the unbeliever in terms of secular pleasure is no less a sticking-point for those who seek to minimize potential loss. Perhaps, therefore, the argument of the Wager is weightiest for those who find their present lives a tasteless burden and who have every reason to aspire after a better life in the beyond.

From the standpoint of the maximization of gain, which is after all Pascal's major concern, it is indeed remarkable how little he offers in the argument of the Wager either to attract or to deter the unbeliever. Partly this was in the inevitable nature of things, but partly also there seems to have been some omission on his part. (To do him justice, we must remember that this thought, like all the others, was left unfinished.) Why is there no mention of Hell, a consideration which might have induced the unbeliever to wish to maximize his gain? Although the love of God for mankind, rather than His retributive anger, is central to Pascal's religious outlook, there can be little doubt that some reference to the eternal suffering of those who reject their Creator would have provided a 'powerful and concurrent' reason for the unbeliever's conversion.

But what, from the standpoint of the maximization of gain, is perhaps most obviously lacking is a recognition of the important semantic and philosophical difference between the proposition that there is a 1:1 chance of there being *something* beyond the grave and, on the other hand, the definition of *precisely what* may exist there. On what basis does Pascal postulate a 1:1 chance of an infinity of *infinitely happy* life to follow our life on earth? At the very most, it could be argued, there is a 1:1 chance of something rather than nothingness in the Hereafter; but infinity? and an infinite existence of infinite happiness? This being so, the unbeliever is quite justified in his refusal to wager. Who knows, after all, whether the afterlife may not be an abode of demons maliciously determined to revile and even torture the spirits of departed humanity, regardless of the quality of the lives they have lived upon this earth? This is a perfectly possible eschatological situation, yet it is not one to which Pascal addresses himself. The Either/Or of the Pascalian Wager (foreshadowing Kierkegaard)[3] is founded upon the alternatives of either blank nothingness or a Hereafter presided over by a God identical, or akin, to the Christian God. It thus becomes clear that Pascal in the Wager is not really speaking to the agnostic so much as to the libertine. In other words, Pascal speaks to his reluctant wagerer from the common position of a shared understanding of the Christian Heaven, even though the one may believe in it whilst the other does not. The argument of the Wager will not convince the agnostic of the specific truth of the Christian doctrine of an afterlife; it may, however, despite its disregard of Hell, incline the Christianly minded libertine – the man of dissolute conduct who feels no binding commitment to Christianity – towards a reassessment of his life in the light of the possible existence of the Christian Hereafter. It may or may not do this; but, so quantitatively expressed, it cannot dispose him to a love of God. As such, it comes perilously close to the Jesuit doctrine of attrition which Pascal had denounced in the tenth of the *Provincial Letters*.

Does not the whole of the argument of the Wager pander, in fact, to man's baseness and cupidity? Do not the blandishments of an infinity of *infinitely happy* life smack of an earthly concept of pleasure? It must be remembered that the argument of the Wager is merely the hinge of Pascal's Apologia, not the door itself, still less what may be beyond. Here, as throughout the Apologia, Pascal would have taken particular care to speak to man in a language he could understand. How otherwise could he have conveyed a sense of celestial pleasures, the pleasures of the 'night of fire', to a dissolute man only capable as yet of appreciating earthly delights? But that the earthly is no substitute for the heavenly, rather the mere shadow of a Reality, Pascal leaves the reader in no doubt. Perhaps, however, he takes insufficient account of the fact that libertines and unbelievers (even if not his puppet libertine manipulated by himself in the argument of the Wager) may calmly and serenely face the prospect of total extinction. Such happiness as is, or has been, theirs on earth they may not wish to be perpetuated into a Beyond. It is, therefore, ultimately on the concept of Hell that Pascal in the Wager should have rested his case. The libertine may otherwise be perfectly content to live his life of pleasure as he chooses, without reference to futurity. But Hell is mentioned only eight times in the New Testament, and even then in uncertain terms; and Pascal was not a firm believer in eschatological punishments. It is perhaps hard for a Jansenist, or one of Jansenist sympathies, to believe that God creates an eternity of torment for those souls whom He chooses not to save.

Regardless of the doctrine of Hell, Pascal's libertine – who admittedly is a docile creature – fully accepts the force of the mathematical argument. The only difficulty which remains, and it is a great one, is that he still cannot *believe*. Although, in other words, as a libertine he accepts the force of the moral argument, as an agnostic he cannot accept the force of Chrstian doctrine. Though the Wager may provide a spur to action, it cannot enforce belief. So far as the faith of the Church is concerned, he remains an outsider. The rest of the argument of the Wager deals with this problem and indeed foreshadows the whole of the second half of the *Thoughts*.

This second half, it will be recalled, was intended to show that 'there is a Redeemer, as is proved by the Scriptures' and, having first made men wish that religion were true, to 'show that it is true'. Having brought men through a heightened awareness of their own nature to the supreme challenge of the Wager, and succeeding in that challenge, Pascal is now on the threshold of defending not natural religion, or religion generally, but the specific truths of Christianity as mediated through the Scriptures and the Church. But although the Scriptures are, in fact, the central subject of Part II of the Apologia, he devotes only half a line to them in the argument of the Wager. His main objective in this latter part of the argument is to dispel the wagerer's 'invincible ignorance'. 'My hands are tied and my lips are sealed', the latter protests; 'I am being forced to wager and I am not free; I am being held fast and I am so made that I cannot believe. What would you have me do then?'

Pascal's reply to this plaintive question has aroused great controversy. He insists that agnostics who have perceived the rational argument but who are still incapable of faith must go through all the notions of belief, behaving 'just as if they did believe, taking holy water, having masses said, etc'.

> That will make you believe quite naturally, and will make you more docile.

This bold pronouncement has frequently caused Pascal to be accused of a kind of religious brainwashing. Regardless of the inherent truth of religious propositions, go to Church often, behave as all the others do whether or not you believe in the various practices – and you will believe! Put in this way, Pascal's argument can be charged with intellectual dishonesty. He amplifies his point in *Thought* 821*:

> For make no mistake about it: we are automata as much as we are minds. And hence it comes about that the instrument of persuasion is not demonstration alone. How few things can be demonstrated? Proofs convince only the mind; custom provides the strongest proofs, and the ones that are the most believed. It inclines the automaton, which leads the mind unwittingly along with it. Who has proved by demonstration that it will be daylight tomorrow or that we shall die, yet what things are more believed in? So it is custom which persuades us of these things. It is custom which makes so many Christians, custom which makes Turks, heathen, trades, soldiers, etc. The faith received at baptism is what Christians have and the heathen have not. In short, we must resort to custom once the mind has seen where truth lies, so as to drench and dye ourselves in that belief which is for ever eluding us, for it is too much trouble to have the proofs always present before our minds. We must acquire an easier belief, which is that of habit . . . When we believe only because of the strength of our conviction whereas the automaton is inclined to believe the opposite, that is not enough. We must therefore make both parts of ourselves believe: the mind by virtue of the reasons which need be seen only once in a lifetime, and the automaton through being accustomed to that, and by not allowing it any inclination to the contrary. *O God, incline my heart . . .*[4]

It is the passions which mislead our reason and, consequently, will still pervert the agnostic even when he has admitted the compelling reasonableness of the Wager. And therefore, if any true perception of life is to be achieved, the passions must be subdued. 'Your inability to believe', Pascal tells the agnostic towards the end of his argument of the Wager,

> stems from your passions. Since reason leads you in that direction and yet you still cannot do so, work hard not at convincing

> yourself by means of multiplying the proofs of God's existence but by subduing your passions . . .

So, in Pascal's philosophy, the heart is the source of two conflicting impulses. It inspires the passions which lead reason astray, and yet it is also at the very foundation of right thinking. For without it, reason will never attain those first principles upon which all reasoning must be based: 'the heart has its reasons of which reason knows nothing' (423*); 'all our reasoning comes down to yielding to feeling' (530*); 'it is the heart which senses God's existence, and not reason' (424*);

> we know the truth not only through our reason but also through our hearts. It is in this latter way that we know first principles; and reason, which has nothing to do with the matter, tries in vain to resist them . . . For knowledge of first principles – space, time, motion, numbers – is as firm as it is of any of those derived from reason, and it is upon such forms of knowledge derived from the heart and instinct that reason has to rely and upon them that it must base all its argument. The heart senses that there are three spatial dimensions and that the quantity of numbers is infinite, then reason goes on to demonstrate that there are no two square numbers one of which is double the other. Principles are sensed, propositions are conclusively proved, and everything is arrived at with certainty, albeit by different methods . . . (110*).

Likewise, just as the heart has a dual function in Pascal's analysis of human nature, so too habit or custom can be variously envisaged. On the one hand, custom is a 'deceptive power' (45*) corrupting and distorting our understanding of the world; yet, on the other, it is that force of habit which settles and stabilizes our true perception of things, once they have been rightly apprehended. Custom is a 'deceptive power' in so far as it leads man to think that there is a rightness and a permanence in human institutions which can, in fact, exist only in the divine. But habit, mastering the *automaton* in man, is necessary to maintain a man's belief in God's existence and providence when once – perhaps only in the briefest moment of insight – these have been assented to by reason and the heart. Thus, the thought on the automaton is very far removed from brainwashing: the Christian religion being the fellowship of a community, how can its truths be understood except through the sacramental life, worship and indeed ritual of that community? *Where two or three are gathered together in My name, there am I in the midst of them.*[5] Only through entrance into that fellowship, and through participation in its ritual life even to the extent of crossing oneself with holy water from the stoup, can one subdue the passions of 'the world, the flesh, and the devil' which draw one away from holiness. The fullness of the religious life comes only through a deeper learning and a deeper feeling; indeed, without that learning and feeling no religious life will come at all. Who (to take a comparable experience) would grow to understand and appreciate the beauties of painting without frequent and at first, perhaps, rather reluctant visits to picture galleries? The Christian message cannot be apprehended by reasoning alone, since it is a religion not based upon the natural unaided reason but upon supernatural *revelation*.

9

The Scriptures

The supernatural revelation on which, and only on which, Christian faith can solidly be established comes to man primarily through the Scriptures, of which little is said in the actual argument of the wager other than that they will enable the wagerer to 'see what the cards are'. But in the total plan of Pascal's Apologia the Scriptures – Old Testament, Apocrypha, New Testament, and even such Judaic writings as the Talmud – would have been of paramount importance. There would thus have been a twofold spiritual nourishment of the converted wagerer: a nourishment of his reason by study of the Bible and other sacred writings; and, through the ritual practices of the Church, a nourishment of his heart by which alone the essential principles of reasoning can be perceived.

The amount of thought and reading which Pascal, even before his death, had put into his sketches for Part II of the Apologia is altogether astounding. He was evidently equipped with as fluent a command of Latin as he had of French, for Latin was still at that time the international language of scientists: his own works on the summing of powers of series of numbers in arithmetical progression, on the factorization of multiples, and on certain aspects of combinatorial analysis, together also with his 'Generation of Conic Sections' and his three *Circular Letters Concerning the Cycloid* were all written in it; from his letter to Fermat dated 29 July 1654 we know that he found it easiest to write about mathematics in Latin (79). It was therefore equally easy for him to read the Scriptures in their standard Latin translation, the Vulgate: a translation which, unlike the standard English Bible, also contained the Apocrypha. But Pascal went much further than this. From *Thought* 970*, in which he quotes from Eusebius's *History of the Church*, it is clear that he also had some mastery of Greek. And, for his understanding of the Talmud (the Rabbinical commentaries upon the Old Testament), he was largely dependent on a book written, in Latin, some four centuries earlier: Raymundus Martini's *Dagger of the Christian Faith Against the Moors and Jews*, first published in France in 1651. This book had been intended to sift what was of genuine relevance to Christianity in the Talmudic writings from what was plainly heretical. In his study of Martini and the original version of the Old Testament Pascal may also have been helped by a working knowledge of the Hebrew language.[1] The last years of his life, which both from Mylon's letter to Huygens in March 1657 (**III 437**) and from Pascal's own letter to Fermat in August 1660 (522) we know were largely taken up with religious studies, must principally have been devoted to meditation on the Bible.

Being (as the Church taught) the inspired word of God, the Bible was the unique, if not quite the sole, medium whereby the truth of the Christian revelation could be imparted: unique in that it was the ancient and traditional receptacle of Christian religious thought, but not quite the sole medium in that the Biblical commentaries of the ancient Fathers of the Church, together with the teachings of that Church as embodied in the decisions of Councils and Popes, refined and perfected the original doctrines. Nothing of course, in Pascal's view, could carry equal weight with the Bible; and, for a layman of his time, he is quite outstanding in the knowledge of the Bible which the *Thoughts* display. As for the ancillary authorities, Pascal considers that Papal decisions – being more consistently in tune with the developing circumstances of the modern world – are at least as worthy of notice and obedience as are the decisions of Councils. Indeed, in so far as the Pope is at the head of the visible unity of the Church, his pronouncements are worthier of notice and obedience. 'France', Pascal observes (604*), perhaps with mingled satisfaction and sorrow, 'is now almost the only place left where one is allowed to say that the council is above the Pope'. But neither Popes nor Councils are more than a pale shadowy reflection of the pristine authority of the Bible.

In his treatment of the Scriptures Pascal reveals tact, wisdom, and ingenuity in presenting familiar truths in a new light. He does not beat his agnostic about the head with the Bible when the latter has not accepted its relevance and authority. Although respectful of the authority of both ancient and contemporary religious teaching, he allows himself no ambivalence in his preference for the traditional; yet, inevitably perhaps, he takes a somewhat ahistorical view of the canon of the Bible. The thirty-nine books which compose the Old Testament, the twenty-seven books of the New, and the fourteen of the Apocrypha are, after all, the product of lengthy deliberations and sometimes fierce disagreements within the councils of the Western Church, most notably (under the guidance of St Augustine and St Jerome) in the fourth and fifth centuries. Pascal, however, despite his reading of Eusebius, always writes of them as something which both is and always was immutably final, fashioned (as it were) by God Himself. His Classical reverence for the past feeds on them rather than on Aristotle and Plato, both of whom however he respects (533*). Once his agnostic has taken the decisive step of the Wager, Pascal seems in no doubt that the Scriptures' binding authority will be accepted.

The suggestions, for example, that a miracle of turning water into wine occurred at Cana, or that five loaves and two small fishes fed five thousand people, or that Jairus's daughter was raised from the dead, will be accepted because they are vouched for by St John, St Mark and St Luke. And indeed, in Part II of the Apologia, miracles are a cornerstone of Pascal's defence of Christianity. 'It would have been no sin', he writes (184*), 'not to have believed in Jesus Christ except for the miracles'. In this respect Pascal is at variance with a great body of Christian opinion, including for example Newman,[2] Malebranche[3] and Leibniz[4] (but not the voluntarist Newton).[5] To Malebranche, presumably disregarding the verbal inerrancy of the Scriptures, miracles would have proved that God did not exist. Leibniz was similarly disturbed by the emphasis which Pascal laid upon the historical aspect of the New Testament. 'I truly consider', he wrote in 1683,[6] 'that God has spoken to us not so much through sacred, civil or even natural history as through and within our minds, through those eternal truths which are quite distinct from the material world': a view not unlike that of Kant.[7] Ninian Smart, recalling the doubts about miracles voiced by Hume in his *Enquiry Concerning Human Understanding* (doubts focusing on probability!),[8] questions whether much value can be attached to miracles as evidence of religious truth;[9] there can, in Smart's view, be no limit to the magnitude of the miracles needing to be performed in order to attest divinity. But this is to overlook the specific value which miracles have for Pascal. For they accord with his profound belief in the Hidden God (and the fact that only some light is shed, but not too much): God, through Jesus and the miracles of Jesus, partially revealing Himself in what, to profane eyes, might seem at times to be a ludic display of divine power.[10]

Alongside the importance attached by Pascal to miracles is the overwhelming significance of prophecy. 'The weightiest of the proofs of Jesus Christ', he writes (335*),[11] 'are the prophecies', which, in his view (328*), 'speak of God not by outward proofs but from an inward and immediate feeling'. Ironically, therefore, the unhistorically minded Pascal finds in history the deepest and most incontrovertible proof of the truth of Jesus: history which in the miracles has occurred, and which in the prophecies has miraculously been *fulfilled*. Yet Jesus is not merely a figure of history. He is Jesus Christ both in time and in eternity, the object of Pascal's mystical contemplation in *The Mystery of Jesus* which now ranks as no. 919* of the *Thoughts* and would perhaps have been given a place of honour in the completed Apologia.

In all these ways, tantalizingly fragmentary as they are, Pascal shows great innovatory skill. Yet perhaps an even greater ingenuity would have been employed in bringing out the peculiarly distinctive features of Christianity. For Pascal, after all, is seeking to prove that Christianity is uniquely true, that there is something about this religion which all other religions lack, and that it therefore not only deserves but also compels our ungrudging allegiance. One of the less attractive features of the *Thoughts* is that they reveal Pascal at his fiercest and least ecumenical. Already in the first part he had strongly condemned other religions, just as he had condemned secular philosophies. His prime target had been Islam, denounced partly for its ferocity, intransigence and widespread success in that through the power of the sword (209*) it had already taken a firm grip on Asia and Africa; but also, no doubt, because this was the religion for which Jesus was not sufficient. Pascal emphasizes the 'difference between Jesus Christ and Mohammed' (209*): 'Mohammed not foretold. Jesus foretold. Mohammed slew, Jesus caused His followers to be slain. Mohammed forbade reading, the Apostles commanded it . . .' Mohammed performed no miracles (321*). Pascal condemns the worldliness of Mohammed's view of Heaven (218*). He denies that the Koran was written by Mohammed (207*): the religious testament of Islam is a fabrication, and it is full of obscurities (218*). Pascal does not deny that there are obscurities and even discrepancies in the Bible, and especially within the Old Testament.[12] But the obscurities in the Koran are of a different order, for they are not tempered and counterbalanced by clarities:

> I admit that [in the Scriptures] there are obscurities as odd as Mohammed's, but some things are admirably clear, with prophecies manifestly fulfilled . . . We must not confuse and treat as equal things which are only alike in their obscurity, and not in the clarity which earns respect for the obscurities (218*).

Mohammed, in short, is 'devoid of authority' (203*).

Pascal's attitude towards the Jewish religion, in Part I of the *Thoughts*, had been slightly less severe. That religion was not entirely *without* authority; but, just as Mohammed had found Jesus insufficient, so the Jews had betrayed Him; and they had betrayed Him out of concupiscence, the worldliness which refused to recognize a king in the carpenter's son entering Jerusalem upon an ass. When Pascal attacks the Jews of Biblical times, which is often, the adjective which most frequently springs to his pen is *charnel* (carnal): the Jews are carnal not in the sexual sense but in the sense that they are concerned with matter rather than with spirit, the flesh rather than the soul. By such a carnal people the Prophet whose *kingdom is not of this world* [13] was bound to be reviled and crucified.

For Pascal, therefore, there is one and only one revelation of religious truth and that is the Christian revelation. He is far from sharing the point of view of Hegel, that 'religion is the knowledge that Spirit has of itself as Spirit'[14]: Spirit manifesting itself in man, and man becoming aware of himself as Spirit. In his defence of Christianity there is no compromise, no tolerance of other religions. He will not accept that all religions contain at least a particle of spiritual insight, a fragmentary but nevertheless precious perception of a truth which can in any case never totally be apprehended. Indeed, it is true – undeniable even by Pascal – that some particle of religious truth must exist in both Judaism and Islam, since the authority of the Old Testament is accepted by Islam, Judaism and Christianity. To all three religions Moses is a prophet; Christianity claims to fulfil Judaism just as Islam purports to fulfil Christianity. For Pascal, however, it is not a question of whether a religion contains a fragment of truth, but whether it bears witness to the *whole* truth. And if one religion bears witness to the whole truth, then (by definition) no other religion can. In Pascal's view, it is Christianity which bears witness to the whole truth; and the Jewish revelation is acceptable only to the extent that it foreshadows the Christian one, although it does not fulfil it.

In his attempt to prove the superiority of the Christian religion, it is essential for Pascal to suggest in what ways and by virtue of what features Christianity excels its rivals. Through his contemplation of human nature, and by the argument of the wager, the libertine (we may assume) has been brought to an acceptance of religious values; but these values are mediated by specific religions: which *faith* should he choose, and why? Clearly (and no apologist of the Christian religion has ever realized it more fully than Pascal), the Person of Christ is at the centre of Christianity; and, for the prospective believer, Christianity will stand or fall by His life and teaching. Pascal, in his awareness – derived from Montaigne[15] – of the omnipresence and supreme influence of custom, is only too conscious of the fact that, once converted to a religious view of the world, his unbeliever will naturally opt for the religion of his country: namely, Christianity. The first objective, therefore, is to make his convert glad of the circumstance that Christianity happens to be the religion of his countrymen. Pascal suggests an ingenious reason why Christianity may be superior to the religions of all the other countries into which his convert might have been born; and though, as the *Thoughts* stand, this reason is provided in Part I, it may safely be assumed that the point would have been developed in the second part of the Apologia, since it is only in Part II, through the fellowship of the Church, that the force of this particular argument could have been appreciated.

'Only the Christian religion', he writes (219*), 'is appropriate for all, being a blend of things external and internal.

> It exalts the people inwardly, and humbles the proud outwardly, and is not perfect without both, for the people must understand the spirit of the letter whilst the clever must submit their spirit to that letter.

Once again (the instances of the technique are innumerable!) Pascal chooses the middle way between extremes: Christianity, he argues, is the perfect blend of outwardness and inwardness, elevating the humble and humbling the proud, inspiring the less intelligent to penetrate the spirit of its message, and forcing the more intelligent to accept the material and contingent form in which it reveals itself. There are 'two errors', he writes (252*): '1. to take everything literally, 2. to take everything spiritually'. It is precisely because the Jews 'took everything literally' that they are described as carnal and are found wanting. Christianity, on the other hand, does not 'take everything spiritually': it interprets in a higher sense what the Jews had understood in a lower, and fulfils spiritually the prophecies which in Judaism had been envisaged materially. Yet it does not lose touch with physical reality: it remains the religion of the Incarnation, the Crucifixion and the Resurrection.

Moreover, not everything in the Old Testament is to be interpreted spiritually. This is the sense of Pascal's warning about the 'two errors'. In the essentials, but not in all particulars, Jesus fulfils the Old Testament: not merely the prophetic statements of those commonly accepted as prophets[16] but also the prophetic insights of the Old Testament generally, and especially the Psalms.[17] In Pascal's words (276*), 'the Old Testament is a cypher', full of *figures* – that is to say, symbols or metaphors[18] – which can be, and were, viewed both literally and metaphorically: for it is of the essence of his defence of Christianity that there is just enough clarity, but not too much. And it is because of the very ambivalence of these metaphors, such as those of the Suffering Servant,[19] the Stone which the builders rejected,[20] the Law,[21] the Temple[22] and the Babylonian Captivity,[23] that the Jews mistook the teaching of Jesus:

> The carnal Jews understood neither the greatness nor the lowliness of the Messiah as foretold in their prophecies (256*).

It is a theme which Pascal proceeds to develop with immense subtlety.

The fact that Jesus is thus foreshadowed in the Old Testament, though the Jews failed to recognize this, impresses Pascal as an argument of incalculable importance. It represents the continuity of God's purposes from the beginning of Creation, and their *perpetuity* (279*-89*) until the end of the world. It expresses the vitality of a tradition extending even beyond the invention of handwriting (282*). Far from detracting from the originality of the mission of Jesus (as some might think today), His symbolic presence in so much of the Old Testament is seen as a reinforcement of His authority. The truth of Jesus, Pascal argues, is a millennial wisdom towards which men from the beginning of time were groping. Jesus is foreshadowed in the Pentateuch (282*, 290*-7*); His genealogical descent from David, and through David from Adam, is neither too conspicuously obvious nor too inconspicuous (236*); this is the purpose of the Book of Ruth (Ruth being an ancestress of David) and of the story of Thamar (304*, 236*). Even the Midrash, the Rabbinical exposition of the Old Testament, bears witness to the truth of Christian teaching: there is 'ample tradition of original sin according to the Jews' (278*). Original sin, as the controversy between St Augustine and Pelagius illustrates, is a doctrine formulated and developed by the Church rather than expressly present in the

Scriptures: just as the Old Testament is fulfilled in the New, so the Rabbinical teaching is fulfilled in the teaching of the Christian Church. The doctrine of Original Sin is no phantasm of the mind devised by St Augustine and others: it is stated by Pascal (278*) to be present in the Scriptures as early as Genesis VIII 21. The fact that it has its roots in the past is a sign of its importance for the future. The partial perceptions of the Old Testament and the Rabbis are fulfilled in the complete perceptions of the Gospels, Epistles and Church Fathers. Moreover, the teaching of the Rabbis has continued into the Christian era; and its greatest Rabbi, Moses Maimonides, has come to accept the authority of Jesus – at least, up to a point. 'Moses Maimonides says that Scripture certainly has two aspects and that the prophets foretold only Jesus Christ' (274*). Pascal would even have gone to those theosophical books highly admired by Moses Maimonides, the Cabbala, in order to extract from them proofs of both Testaments (274*). Not that he shared the presupposition of the Cabbalists that every passage in the Old Testament is mysteriously symbolic; but somehow he hoped to prove that they too, although extreme in that they detected more light than there really was, threw light nevertheless on the ways in which the New Testament fulfils the Old.

But, it may be objected, is there in fact enough light for the unbeliever to be convinced – or so much darkness that only the most perceptive can avoid being deceived? To illuminate his exposition of the New Testament still further, Pascal proposes his argument of the Three Orders (308*). These, the Order of bodies, the Order of minds, and the Order of charity, are to be conceived as being discontinuous, like three wavelengths operating independently of each other and never coming into direct contact. The Order of bodies is that of physical strength and brute force, worldly power and secular authority. The Order of minds is that of intellectual and imaginative power. The Order of charity is 'the holiness of the heart's affections', ἀγάπη, *caritas*, the love of one's fellow men. Those who are rich in intellect may be lacking in charity; those who are physically weak (like Pascal himself!) may be kings of the intellect; men of lowly birth may be giants of charity. The carpenter's baby son born in a manger in Bethlehem may be the incarnate Logos, *the Lamb slain from the foundation of the world*[24] (259*). In this way Pascal accounts for the fact that ancient historians writing of the Holy Land at the beginning of the Christian era pay little[25] attention to the story of Jesus. There are references to Him in Josephus,[26] Pliny,[27] Tacitus[28] and Suetonius.[29] Pascal was almost certainly unaware of the full extent of these references, but his point is still substantially valid: Jesus, he writes (300*),

> in such obscurity (according to what the world calls obscurity) that historians writing only of important political events scarcely noticed Him.

Nor did Jesus purport to be a man of great intellect:

> Jesus Christ and St Paul possess the order of charity, not of the mind, for they wished to humble, and not to teach (298*).

The message of Jesus is addressed essentially to the heart. The Jews mistook this message, because they looked for literalness, not spirituality. Respectful of the order of bodies, not that of the heart, they failed to recognize the Messiah in a man who would not fight for the worldly deliverance of his people from the Romans. Yet, Pascal argues, all the essential features of His Messiahship had been foretold by the prophets.

The continuity of this revelation is, however, only one of the ways in which it compels respect, and even reverence, from the order of the heart. The other is its miraculous nature. For the mere continuity of a revelation, and its fulfilment some hundreds of years later, do not in themselves constitute proof of its sovereign authority. The fact that it has been a continuously miraculous revelation – from the miraculous happenings recorded in the Old Testament[30] (486*) to the miracles of Jesus Himself – is a different matter. 'It would have been no sin not to have believed in Jesus Christ except for the miracles' (184*). 'It is not possible to have reasonable grounds for not believing in miracles' (568*). The importance of miracles impressed him so much that he seems to have wished to devote to them a whole section of the second part of his Apologia.[31] No doubt they loomed so large in his consciousness because of the miraculous cure allegedly wrought upon his niece Marguerite Périer, by a thorn from the Crown of Thorns, at Port-Royal de Paris on 24 March 1656 – less than a year, it would seem,[32] before he recorded some of his thoughts on miracles (e.g., 901*, 903*). These are, he believes, the 'signs and wonders'[33] which compel belief in the divinity of Jesus.[34] They attest the divinity of the revelation of which Judaism was the source and Christianity the fulfilment:

> Jesus Christ performed miracles, so subsequently did the apostles, and the early saints in great numbers, because, since the prophecies were not yet fulfilled, and were being fulfilled by them, there was no witness except that of miracles . . . Not until He had died, risen again, and converted the nations, were all things fulfilled and therefore miracles were needed throughout that time. Now there is no more need of miracles against the Jews, for the fulfilment of the prophecies is a continuing miracle (180*).

Islam, therefore, is excluded from Pascal's continuity of miraculous revelation: the 'miracles against the Jews' ceased with the immediate end of Jesus's mission and the end of those additional miracles recorded in the Acts of the Apostles.[35] The prophecies being entirely fulfilled in Jesus, and the miracles having ended, this was the definitive revelation. Islam, it should be added, though it abounds in stories of subordinate miracles allegedly performed by minor Islamic saints, does not regard Mohammed as a worker of miracles. On the contrary, as C.J. Wright remarks,[36] Mohammed 'seems to suggest that it was no part of his mission to perform signs and wonders'. Christianity, therefore, is not only a fulfilment of prophecies but a consummation of miracles. 'The fulfilment of the prophecies is a continuing miracle'. And so, within the life of the Church, but against the Jesuits (859*, 877*), no longer *against the Jews*, miracles can continue to occur; as, Pascal believed, the miraculous recovery of his own niece and goddaughter demonstrated.

That Pascal, as a scientist, should have believed in the disturbance of the natural physical world through miracles need not surprise us. Religion, in his view, was a receptacle of supernatural power. God could upset the order of nature in order to attest a profound spiritual truth. Nor, amongst the thinkers of his time, is Pascal by any means alone in his acceptance of miracles. Locke, publicly at least, never doubted their historical authenticity. 'Where such supernatural events', he wrote,[37] 'are suitable to ends aimed at by him, who has the power to change the course of nature, there, under such circumstances, they may be the fitter to procure belief, by how much the more they are beyond or contrary to ordinary observation'. To Joseph Butler, writing his *Analogy of Religion* eighty years after Pascal's work on the *Thoughts*, miracles (with the fulfilment of prophecies) were the 'direct and fundamental proofs'[38] of the truth of Christianity. Berkeley also admitted their genuineness as historical events.[39]

Indeed, the thinkers of the eighteenth century were more disposed to admit miracles than were their predecessors in the seventeenth: a fact perhaps connected with the remarkable flowering of the natural sciences in Pascal's time. Pascal, on this analysis, exhibits a rare blend of faith both in the natural and the supernatural, which (in his judgment), far from being incompatible, complemented each other. Nevertheless, it is agreed even by his most favourable critics that he envisaged the miraculous manifestations of the supernatural with a degree of credulity unbefitting either a natural scientist or a historian. 'Even his clear mind', one of these critics has written,[40] 'was not proof against dogmatic presuppositions in regard to ecclesiastical miracles which no historical critic would now accept', citing in evidence of this Pascal's belief (a commonly held one, but adopted by him without reservation) that it was by a miracle that the Empress Helena recognized the True Cross of Jesus's Crucifixion: miraculously discerning which was His cross and which were the thieves' (901*).

Pascal is much more historically sensitive in his treatment of the transmission of the Gospel story. What, after all, is the evidence for Jesus's mission? Even before the scientific acceptability of miracles is discussed, where is the historical evidence that they were supposed to have occurred? Pascal does not flinch from the fact that the testimony to the historical Jesus in Classical literature is fairly slight: e.g., Tacitus: 'Christus . . . suffered the extreme penalty during the reign of Tiberius at the hands of one of our procurators, Pontius Pilatus'; or Josephus: 'Jesus, a wise man, if it be permissible to call him a man, for he was a doer of wonderful works, a teacher of such men as receive the truth with pleasure . . . This was Christ . . .' For our knowledge of the life, death and resurrection of Jesus, His teaching, miracles, and fulfilment of prophecies, we are very largely dependent upon the Apostles. The four Gospels, together with the Pauline Epistles (notably the Apostle Paul's account of the institution of the Eucharist),[41] are almost the only record we possess of Jesus's mission; they are certainly the only detailed record. What, then, if it were all a fabrication, a conspiracy by the Apostles to deceive their followers and posterity?

Pascal devotes several thoughts, and they are important ones, to what he terms the hypothesis of the 'deceitful Apostles' (310*, 322*, 457*).[42] Is it, he asks, really credible that after the death and Ascension of Jesus these twelve men (including Barnabas) could deliberately have set out to perpetrate a huge confidence trick?

> The Apostles were either deceived or deceivers. Either supposition is difficult, for it is not possible to imagine that a man has risen from the dead. Whilst Jesus was with them, He could sustain them; but afterwards, if He did not appear to them, who made them act? (322*).
>
> The hypothesis that the Apostles were deceitful is quite preposterous. Follow it through and imagine these twelve men meeting after Jesus's death and conspiring to say that He had risen from the dead. This means attacking all the powers that be. The human heart is singularly susceptible to fickleness, change, promises and bribery. One of them had only to deny his story under these inducements, or still more because of possible imprisonment, tortures and death, and they would all have been done for . . . (310*).

This, to Pascal, is the clinching argument as to the authenticity of the miraculous and prophetic revelation. It is one which to Laplace appeared doubtful on mathematical grounds;[43] but nevertheless it has probably made sound sense to the hearts of many of Pascal's readers, even if it failed to convince Laplace's intellect. It is, in fact, the cornerstone of that Scriptural nourishment of the newly converted believer which, as the thought on the Wager insists, must proceed in harmony with his liturgical nourishment and the subduing of the *automaton* within the heart of passionate man.

Thus Pascal brings, or hopes to bring, his wagerer round to an acceptance of the proposition ($p \approx 1$, $s = 0$, where $x = \infty$) that the probability of eternal salvation is no mere probability but a certainty, and that the stake has amounted to *nothing*. At the very outset of the thought on the Wager he had suggested that the stake was nothing, whilst the probability of eternal salvation was 1:1. This the agnostic had refused to accept. The stake, which to him amounted to the giving-up of his worldly pleasures, had seemed very large indeed for so uncertain an outcome. Now, however, with the subduing of the *automaton* and in the deepening knowledge of the Scriptures, his corporate life within the fellowship of the Church will presumably draw him away from the world; and he will not repine. Pascal's final formulation of the Wager is, of course, no more than a figure of mathematical rhetoric; but it is a powerful one and, for those who have been convinced, embodies a deep spiritual truth.

The argument of the Wager is not only a hinge, or a bridge, leading into Part II of the Apologia. It extends across the whole length of the second Part, since the evidences of liturgy and the Scriptures are part and parcel of the deepening religious awareness. If liturgy, miracles and prophecies do their work, then what the wagerer has staked in the way of a finite life will indeed seem *nothing* to him in comparison with what he has gained.

Pascal, unlike Leibniz, does not attempt a total theodicy in that he does not explore the workings of Divine Providence from a strictly philosophical angle. He does not consider the difficulty that, if Christianity is the final and unique revelation of religious truth, God might seem neglectful of the destinies of those born before Christ. He has little to say about morality,[44] and is ambiguous concerning the possibility for all men to redeem themselves by means of the Wager (236*). His principal contributions to Christian apologetics are that he relentlessly brings man face to face with the fact of his own mortality, compels him to reconsider his own human nature, and places the Personality of Jesus (where it ought rightly to be) at the heart of Christianity. His is a supernatural, not a natural religion. He never wearies of stressing his aversion to deism (449* etc). Kant, in his *Critique of Pure Reason,* was perhaps right to claim that one should 'deny that the deist has any belief in God at all, merely leaving open to him the affirmation of the existence of a Primal Being, or Supreme Cause of all things'.[45] Long before Kant, Pascal held the same view of deism. 'I cannot forgive Descartes', he is alleged by Marguerite Périer to have said (41).[46] 'In his whole philosophy he would like to do without God; but he could not help allowing him a flick of the fingers to set the world in motion; after which he had no further use for God'. Likewise, Voltaire, usually so harsh in his haunted condemnation of Pascal, can scarcely be absolved from inconsistency when he demands that we should believe in 'a God of rewards and punishments'[47] whilst denying to that God any greater involvement in the lives of His creation than a Cartesian 'flick of the fingers' – or a judicial pronouncement in the Hereafter: he himself has confessed, with superb irony, that 'if God did not exist, it would be necessary to invent him'. Pascal's God of self-abasement and incarnation has not been invented as a convenient philosophical fiction. He is a fact of history as well as of eternity, symbolic of all the prophetic history that has preceded Him and of all the sufferings in human history until the end of time.

Knowing God without knowing our own wretchedness makes for pride.
Knowing our own wretchedness without knowing God makes for despair.
Knowing Jesus Christ strikes the balance because He shows us both God and our own wretchedness (192*).

10

Pascal, 'Devout Geometrician'[1]

In the intervals of jotting down his *Thoughts* Pascal continued his mathematical work. He was about to make some of his most fruitful mathematical discoveries. The infinitesimal calculus was the direct consequence of these. Leibniz's work on this calculus was inspired by Pascal's treatises on circular arcs and the sines of quarter-circles. Moreover, Pascal's discoveries also anticipated the integral calculus which Newton was to devise in 1665–6, despite their lack of a generalized (fluxional) method.

Early in 1657 we are given a valuable insight into Pascal's outlook and way of life by Claude Mylon, the advocate and dilettante mathematician who may well have shared with François Le Pailleur the chairmanship of Mersenne's scientific circle after the latter's death. 'Although', wrote Mylon, 'it is very difficult to contact M. Pascal and he has withdrawn into complete retirement from the world so as to give himself up entirely to his devotions, he has not lost sight of mathematics. Whenever M. de Carcavi can seek him out and put some question to him, he does not refuse to give him the solution, especially in the matter of games of chance which he was the first to bring to notice' (**III 437**).[2]

One such question is mentioned in the letter from Carcavi to Huygens dated 28 September 1656.[3] Pascal had recently put it to Fermat, doubting however whether even he would be able to solve it. 'Incomparably more difficult than all the others' (in Pascal's judgment: **III 437**),[4] it concerned the respective probabilities in a game of chance in which two players – each of them having three dice, the one aiming to throw 11s, the other 14s, a point being scored whenever the player's particular number is thrown, but scored wherever possible by the subtraction of a point from one's opponent – compete with each other to become the first player to score twelve points. What, then, were the longest odds that could be laid against the possibility of player '14' winning, in the event of player '11' gaining the initial advantage? With regard to Fermat, Pascal's doubtfulness was misplaced. The correct solution to the problem was returned from Toulouse almost immediately: that, expressed in integers, the odds were 1,156 to 1, whereas 1,157 to 1 would be excessive. Then, just to show that he too possessed his own method of solution, Pascal asked Carcavi to write to Fermat on his behalf supplying the precise figures. These, reduced to their lowest terms, made the improper fraction:[5]

$$\frac{150,094,635,296,999,122}{129,746,337,890,625} \text{ to } 1.$$

Carcavi had by now become Pascal's closest scientific colleague, a more intimate helper and associate than Claude Mylon who remarks that in March 1657 Pascal and the former diplomat were often to be seen together at church and in business circles, neither of which areas of life he was in the habit of frequenting.

Marguerite Périer provides two of the reasons for her uncle's continuing devotion to mathematical research (40). One evening, perhaps not long after the completion of the *Provincial Letters*, Pascal suffered an excruciating attack of toothache. He went to bed but could not sleep, and the pain only increased. To alleviate his discomfort by turning his mind away from it, he began to think about the problem of cycloids which Mersenne had first become aware of even before he himself was born and had finally made public after twenty-nine years' meditation on the subject (194). Its solution, with all the relevant proofs, came to Pascal in a flash – and he realized that the toothache had gone. The Duc de Roannez, who had been with him the previous evening and was anxious about his state of health, returned the next morning to see how he was. Instead of finding his friend in pain, he found that he had just triumphantly solved a baffling mathematical problem. To Pascal mathematics had simply been a way of passing as cheerfully as possible a night made sleepless by toothache. Roannez thought otherwise. He urged Pascal not to neglect such intellectual attainments, nor to keep them in a compartment all to themselves; for, quite apart from its therapeutic value, Pascal's mathematical skill could be a powerful weapon in future religious controversy.[6]

This was not bravado on the duke's part. His attitude in no way resembled that of the promoter of a prizefighter keen to take on all comers and thus to assert his own particular, if not general, superiority. Like so many of his contemporaries, Roannez almost certainly felt that no philosophical issue could properly be settled by anyone lacking mathematical ability: Leibniz has expressed the same point of view.[7]

Pascal's work on cycloids occupied him fairly continuously between June 1658 and February 1659. It was his last major contribution to mathematics, and indeed his last outstanding intellectual achievement of any kind outside the sphere of religion. It led him away from the solitary and austere spiritual meditations on which he was also engaged in the intervals of his mathematics, and back into the world. Even, however, after the deep mystical experience of the 'second conversion', and in the midst of his work on the *Thoughts*, his scientific worldliness never lost that tinge of aggressive superiority which had characterized his much earlier disputes with Jacques Forton and Étienne Noël. Pascal was not alone in such zeal for fame and pre-eminence; scientific bickering inspired to some extent by personal vanity was a shortcoming of the whole seventeenth century; nevertheless, it is strange to find it in the author of *The Mystery of Jesus* and the *Comparison between Christians of Early Times and Those of Today*, who was a man of submissive piety intensely aware of spiritual values. The desire to issue challenges, to score a point over one's rivals, and to establish a priority in scientific discovery brought out the worst in Pascal's character; yet it was little different from the animus of Roberval's disputes with Descartes twenty years previously, or from that of the dispute between Newton and Leibniz and their followers which, beginning thirteen years after Pascal's death,[8] continued until 1724. By the conventions of the time there was nothing unusual about Pascal's prize questions on the cycloid: in the previous year (1657) Fermat had issued two challenges to his fellow mathematicians, one of them based on his study of the indeterminate equation $x^2 - q = my^2$ for non-square m. Likewise, Huygens was to test Leibniz with a challenge in 1672, that of finding the sum of the infinite series of reciprocal triangular numbers. Moreover, in 1674 Leibniz was to be challenged, and defeated, by Ozanam's 'six-square problem': that of finding three numbers x, y and z such that $x - y$, $y - z$, $x - z$, $x^2 - y^2$, $y^2 - z^2$ and $x^2 - z^2$ are all squares.[9]

Pascal, in his investigation of the properties of the cycloid, was mainly concerned with four questions: the surface area of any geometrical figure bounded by a cycloidal arc (the so-called *quadrature* of the cycloid); the centre of gravity of the segment so produced; the volumes of various solids of revolution generated by these segments (the *cubature* of the cycloid); and the centres of gravity of these and other solids. Imagine a simple, not a prolate or curtate, cycloid – the curve described by any point on a circumference revolving along a straight line at a uniform speed, for instance by a wheel rotating along a road. First, what is the surface area of the segment bounded by a section of the cycloid, its base, and its symmetry axis? Next, what is the volume generated by this segment as it revolves around its base? What also is the volume of the solid of revolution generated by the segment of the cycloid turning around its axis? Finally, and most important of all, what are the centres of gravity of each of these solids of revolution (as generated by rotation around the axis and the base respectively)? and what are the centres of gravity of each of the halves of these two solids of revolution if we suppose the semisolids to be bisected by midplanes?[10]

The cycloid, with the straight line and the circle, is – as Pascal himself has pointed out (194) – one of the commonest lines within the whole natural world; yet it was never investigated by the Ancients, not even by Archimedes. Well before 1658, however, various geometricians had given it some of their attention. Except perhaps for Nicholas of Cusa (in the fifteenth century), Galileo was the first geometrician to have even conceived of its existence;[11] but, unlike Nicholas, he had conceived of it as a practical rather than a theoretical problem. In the closing years of the sixteenth century he became aware of the cycloid whilst constructing the arches of a bridge, and set about calculating its surface area by weighing a plate of lead of uniform thickness having the shape of a plane bounded by a cycloid: he thus concluded that the area of the cycloid seemed to be about three times as great as that of its generating circle, but could not theoretically establish his findings.

With the remarkable flowering of geometrical studies in the seventeenth century, notable advances in the study of the cycloid curve came rapidly. Mersenne, always the spur to others' achievements rather than the man of achievement in his own right, noticed the cycloid in 1615. He spent much time in vainly investigating its properties and (more fruitfully!) in corresponding with his friends about it.[12] Fermat, without publishing his findings, provided theoretical confirmation of Galileo's. Torricelli considered the point describing a cycloid as being endowed with two simultaneous motions, the one uniform and the other varying; he showed that, given a curve for distance as a function of time, the tangent at any point of that curve forms with the time axis an angle the tangent of which measures the speed of the moving object.[13] And Roberval, in the course of more than twenty years' work on the properties of the cycloid, proved independently of Fermat that the area of a cycloid is exactly three times that of its generating circle, and moreover determined the centres of gravity both of an arc of a cycloid and of a segment of the arc, together with the volumes generated by the rotation of the cycloid around both its base and its axis. Thus, as early perhaps as 1652, Roberval had come very close to the work on which Pascal was about to embark. He had indeed already solved the majority of the problems which Pascal was to issue as a challenge to his fellow mathematicians. But until mid-1658 Pascal seems to have been genuinely unaware of most if not all of Roberval's work on the centres of gravity, and the volumes of solids of revolution, of a cycloid (**VIII 159 n. 1**) – very little of which had been made public either in writing or even presumably in conversation with Roberval's friends.

It would be unjust to detect plagiarism in any of these great scientific men, although Roberval and Descartes frequently bandied that accusation about. Scientific discovery appears sometimes to proceed by a strange communal telepathy, as seemingly independent researchers, often in far-flung countries, arrive at a more or less identical result at about the same time: the phenomenon, so fruitfully investigated by Popper and Kuhn, of which the parallel research work on the double-helix structure of the DNA molecule has been the foremost recent example.[14] Thus, about 1635, Cavalieri, Roberval, Fermat and Descartes were all at work on the geometry of indivisibles, and all constructing tangents to cycloids;[15] likewise, thirty or forty years later, Newton and Leibniz defined independent algorithms of the integral and differential calculus.[16] Whether Pascal had resumed his visits to the scientific Academy now presided over by Henri-Louis Habert de Montmor, and meeting each week at the latter's home in the Rue Sainte-Avoye (close to the Cloître Saint-Merri), we do not know; certainly, however, speculation concerning the cycloid and related problems was rife. At the time of his most intense absorption in the natural sciences, in 1653–4, Pascal may have become familiar with Torricelli's discoveries. He does not seem to have been aware of Galileo's work.[17] Cavalieri's method of indivisibles, of 1629,[18] actually foreshadows the quadrature of the cycloid, being based on the hypothesis that the volume of a curvilinear solid may be calculated by dividing that solid into an indefinite number of no longer divisible plane sections. Descartes, in Book II of his *Geometry*,[19] gave the construction of the tangent to a cycloid. Fermat must have mentioned to Pascal his own work on the ratio of the surfaces of cycloids to those of circles. And Roberval, who seems in his volatile way to have been friendlier with Pascal than with any other scientific colleague, may also have mentioned to him his own preoccupation with the volumes of solids of revolution. When, therefore, Pascal returned briefly but dazzlingly into the scientific arena, it is scarcely or not at all surprising that it should have been to explore the quadrature, cubature, and centres of gravity of cycloid curves. His sudden intervention might also have come as little surprise to his contemporaries, despite his comparative withdrawal from the world, except that for five months or more[20] they were to remain unaware of the identity of the Anonymous scientific challenger.

Not only Marguerite Périer but also her mother Gilberte have described the remarkable circumstances in which Pascal happened to make his discoveries. A night made sleepless by toothache produced the intuition from which a succession of other insights rapidly followed: 'the first [thought] was followed by a second', writes Gilberte (19; cf. 40), 'and the second by a third, and finally by a multitude of thoughts each succeeding one another; these revealed to him, almost involuntarily, the proof of the properties of the cycloid, by which he himself was surprised'. These proofs produced a cure for the toothache, which, writes Marguerite Périer (40), was 'all that he desired of them'. Whether it was, in fact, a cure, or merely a distraction of Pascal's mind from the peculiarly distressing symptoms whilst nature followed her own course, is another matter. Certainly, it seems, without that bout of toothache Pascal would never have returned with so much zest, intensity and ingenuity to mathematics. For it was mathematics more than any other subject, secular or religious, which could entrance him even if only in short bursts of fierce concentration that absorbed and co-ordinated all his intellectual energies. He appears at first to have had no intention of writing anything about the cycloid, preferring to let his discoveries lapse into oblivion: he 'set no store by his discovery, considering it as vain and useless, and not wishing to interrupt whatever assiduity he had for his work concerning religion' (**VII 339–40**). However, with a view to combating 'atheists and libertines' (**VII 340**; cf. 40, also 39), Pascal set about recording his thoughts on the cycloid, and issuing his challenge.

When the writing of the *First Circular Letter Concerning the Cycloid* began, inspiration came rapidly. 'It is incredible', writes Gilberte (20), 'how quickly he dashed that down on paper. For he wrote as fast as his hand would go, and had finished it within a very few days; he never made any copy of what he wrote, but handed the sheets over as he went along. They were also printing another thing of his[21] which he was handing them in the same way, as he went along, and so simultaneously he was supplying the printers with two different things'. As soon as the *First Circular Letter* was printed (June 1658), Pascal and his friends – Carcavi, Mylon (**VIII 3, 17**) and perhaps Auzout – despatched it throughout France and abroad to any mathematicians of their acquaintance who might be capable of resolving the various problems. Wallis,[22] Fermat, Sir Christopher Wren, Schooten and the Flemish canon René de Sluse received copies, as also did Huygens, Michelangelo Ricci and Leibniz. The problems at issue, divided under six heads, were those (previously mentioned) concerning the surface area, volumes, and centres of gravity of a cycloid's segments and of its solids of revolution. Pascal's friend, the former diplomat Pierre de Carcavi was to be chairman of a panel of judges in the competition, and mathematicians were to send him their answers to the six questions posed by the Anonymous challenger. Competitors would be allowed three whole months in which to produce their answers, and the deadline for the certified despatch of entries was 1 October 1658.

Carcavi and his panel would verify the accuracy of the various submissions; and, if appropriate, a first and a second prize would be awarded. The first prize would only be awarded to a competitor who successfully resolved all six questions, whilst the second prize would go to the best runner-up. The amounts of the prizes were forty and twenty *pistoles* respectively, a *pistole* being a *louis d'or* equivalent in value to ten francs.[23] These sums were, or appeared to be, offered by Pascal himself, who deposited them on trust with Carcavi. Roannez, however, may have helped his friend to produce the sixty *pistoles*; he may even, perhaps, have donated the full amount himself. It was only fitting that he should not have to lose financially by his action, which could bring nothing but renown to the Christian religion.

In such an elaborate series of arrangements it was also important that on every side justice should be seen to be done. The prestige which Christianity stood, indirectly, to gain from the competition would be tarnished by any suggestion of unfairness or impropriety.

Yet within weeks of the *First Circular Letter* being distributed throughout France, England, the Low Countries, Germany and Italy, a serious hitch in the arrangements had nevertheless occurred. Pascal suddenly realized (perhaps some would-be competitor brought it to his attention) that of the six questions involved in the challenge, four – relating to the surface area and centre of gravity of segments, and the volumes of solids of revolution, of a cycloid – had already been successfully resolved by Roberval. He therefore instructed Carcavi not to take these four into consideration, insisting that the competition should be judged on answers to the last two (concerning the centres of gravity of solids of revolution), and on those alone. He also, in July 1658, issued a *Second Circular Letter Concerning the Cycloid* (written, as was the first, in Latin) in which he clarified yet another ambiguity: the cycloid at issue was, he explained (**VIII 17–18**), a simple one, not curtate or prolate. The deadline of 1 October for the submission of entries now only applied to the results, and not to the demonstrations.

It was a competition which attracted many more fascinated bystanders than it did competitors. Whereas only two mathematicians were bold or intelligent enough to compete, many followed the proceedings with passionate interest, toying perhaps with partial solutions or ancillary lines of investigation. As early as 6 July 1658, within a month of the start of the competition, Sluse wrote to inform Pascal that he had long since discovered how to calculate the area bounded by a cycloidal arc (**VIII 12**); as for the volumes and the centres of gravity (he writes in a further letter, dated 2 August 1658 : **VIII 116**), these problems seemed so enormously difficult to him that he doubted whether he would have either the time or the skill to resolve them. Roberval's achievements in this nearly impossible field of research were at this time still unknown to Sluse, and (it seems) had only just been discovered by Pascal.

Huygens, like Sluse, came up with a partial solution of the problems[24] but did not submit an entry: the centres of gravity of a cycloid's solids of revolution, and the volume of the solid generated by its rotation around its axis, both eluded him. Another mathematician who made some headway, but without going far enough, was Ricci. And the twenty-five-year-old Christopher Wren made no headway at all with the specific problems, but alighted on the discovery (unfortunately irrelevant to the competition) that the length of the arc of a cycloid is four times that of the diameter of its generating circle. This first ever rectification of the cycloidal arc delighted Pascal. 'Nothing', he later wrote (198), 'is finer than what has been sent in by Mr Wren; for quite apart from the fine method which he provides for calculating the plane surface of the cycloid, he has provided the method of comparison between the actual curve and its proportionate relationship to the straight line' (**VIII 135–6**). Wren, however, as Pascal goes on to remark (198), had merely submitted the 'enunciation without proof'. It was left to John Wallis to publish Wren's method of arriving at his result.[25]

The two actual candidates in the competition set up by Pascal, John Wallis and the French Jesuit priest Antoine de La Loubère, each came out of the contest unsuccessful, full of grievances, and (especially in the Jesuit's case) sorely bruised. Yet both men were more than competent in their line of study (Wallis indeed, with his work on the arithmetic of infinities and on the imaginary roots of quadratic equations, ranks amongst the foremost of English mathematicians); and both were acting out of good faith. Wallis's first letter to Carcavi, dated 19 August 1658, was little more than the outline of a method of integral calculus illustrated by a certain number of calculations which were themselves inaccurate. Wallis also complained on 19 August of the closeness of the deadline, and of the unjustifiably short amount of time allowed for the solution of the two questions in particular which were extremely hard.

This provoked Pascal's *Third Circular Letter Concerning the Cycloid*, published in Latin[26] on 9 October 1658, by which time the deadline for the despatch of entries to the competition had just expired. He indignantly refutes Wallis's suggestion that eight or even twelve months might have been allowed. Even those, he says, who might appear to have won, by lodging their entries with Carcavi on or before 1 October, may not in fact have done so: correct solutions bearing an earlier date might still reach Carcavi after the first of the month, certified 'by the signature of the mayors and officials of some town that people have scarcely heard of, in furthest Muscovy, Tartary, Cochin China or Japan . . . ' (**VIII 161**) Thus it was not only a competition against one's own imperceptiveness, but a race against time in which every day counted. Meanwhile Wallis, in a further letter to Carcavi dated 3 September 1658 (**VIII 233**), had pointed out and rectified his worst errors of calculation. In a third letter to Carcavi (**VIII 233**) still more corrections followed. John Wallis clearly had no claim to any prize.

The only other contestant, Antoine de La Loubère, actually disclaimed the intention of competing for either of the awards (208;

VIII 158). He had begun, optimistically enough, with a solution of the first three of the cycloid problems: these he had sent in to Carcavi by the end of July (**VIII 24**). Dropping his mask of anonymity, Pascal replied on 4 September 1658 (**VIII 121–4**) that answers to three questions were not enough. In a furious race against a deadline La Loubère somehow solved the fourth (**VIII 123**): Pascal replied on 11 September (519) that that had already been solved by Roberval (who was not, however, a party to the competition). There remained the fifth and sixth questions, on which La Loubère must now have worked more furiously than ever (**VIII 157**). This was the crux of the challenge, which no one except La Loubère came anywhere near to solving.

Even before Carcavi eventually received La Loubère's answers to the fifth and sixth questions, which were despatched on 15 September, their author had discovered that they contained serious mathematical errors. He had followed up his initial letter concerning centres of gravity with a corrective. A spate of letters from La Loubère rained in on Carcavi during the months of September, October and November explaining that he no longer wished to be considered a competitor because of the faultiness of his calculations – but, perhaps because by now he was hopelessly out of his depth, not however troubling to send in any reworkings. It was still no mean achievement for a mathematician of secondary status to have done what others of the first rank had so signally failed to do: tackling and mastering the difficulties of the first three cycloid questions within ten days (**VIII 24**).

Moreover, the mathematician of the second rank was in the right, and the first-rate one sadly misguided, when La Loubère was taken to task by Pascal (**VIII 123**) for seeking the formula for the quadrature of the cyclocylindroid, having (so he alleged) already found it. But La Loubère plainly saw that over and above the specific case of the cyclocylindroid extended over a plane surface in which (when the radius of the sphere equals the diameter of the circle at the cylinder's base) the cyclocylindrical curve is equivalent to the so-called *companion of the cycloid*, there is also the further and more problematical one: that of the intersection of any right cylinder with a sphere whose centre lies on the cylinder's surface. He remarks in a letter to Carcavi, admittedly a few months later (**VIII 123**), how interesting it would be to square the cyclocylindroid that is not extended over a plane surface. The distinction between primary and generalized cyclocylindroids was not one of which

Pascal himself seemed aware. In La Loubère's letters, Pascal mockingly writes (198), 'he speaks of the quadrature of this figure which he calls a cyclocylindroid, as if it were something very far removed from his knowledge and something which he would greatly have liked to know about'. Certainly, if Pascal was aware of the distinction, he did not speak of his scientific colleague with the praise and recognition due to a sincere and earnest fellow seeker after truth.

This disparaging reference to La Loubère, and numerous others of a like kind, occurred in Pascal's *History of the Cycloid* (194–200), which was published on 10 October 1658 only ten days after the deadline for the despatch of entries to the cycloid competition, and the day immediately following the *Third Circular Letter*. In his *History of the Cycloid*, very probably written in collaboration with Roberval in view of his friendship with the latter and Roberval's wide experience of this field of geometry, Pascal sets out the history of mathematical research into the cycloid from the early seventeenth century up to his own day. Towards the end of the eight-page pamphlet he touches on his own competition, commenting unfavourably on the submissions received from La Loubère and Wallis. But despite his grievances with both these men, and theirs with him, it was not his remarks on his rivals' entries and lack of success which roused Wallis to a fury and made the *History of the Cycloid* the most controversial scientific paper Pascal ever wrote – even more controversial, in a sense, than were the *Provincial Letters*. What made the *History of the Cycloid* so disturbingly provocative was its treatment of Torricelli.

Pascal begins (somewhat inaccurately, in view of Galileo's practical calculations some twenty years earlier still) by observing that Mersenne in 1615 seems to have been the first scientist to notice the cycloid's existence (194). 'He then tried to fathom its nature and properties, but could not fathom them. He had a quite special talent for formulating fine questions, in which, perhaps, he had no equal: but even though he was not equally fortunate in resolving them, and although it is in this that all the honour really lies, it is true nevertheless that we are under an obligation to him and that he has provided the occasion for several fine discoveries, which perhaps would never have been made if he had not inspired scientists in those directions' (194). Having failed to fathom the secrets of the cycloid himself, Mersenne put the problem 'to all those in Europe whom he thought capable of it, including Galileo. But none had any success with it, and all despaired' (194). Several years, it is said, then elapsed before Roberval managed to penetrate some of the cycloid's mysteries, discovering in particular (with the help of Cavalieri's method of indivisibles, some knowledge of which was available from 1629 onwards)[27] that the area of the cycloid is three times as great as that of its generating circle: that was in 1634 (194). No sooner had Roberval made his discovery than he passed on the good news to Mersenne – and devised a competition! The latter, with his usual obliging officiousness, put all the arrangements in hand and wrote throughout Europe to all the mathematicians skilled enough to be invited to compete, allowing them a year in which to come up with their findings. The object of this original cycloid competition was to determine the ratio of the area of a cycloid to that of its generating circle. Pascal at this time was aged eleven, and sufficiently precocious to have embarked on a Treatise on Sounds. More than twelve months then elapsed, during which time no competitor even entered the lists. Mersenne finally wrote yet again to all concerned, announcing Roberval's solution. At this point, Pascal continues (195), 'two competitors emerged, providing the demonstration': one was from Fermat, and the other from Descartes. 'Their solutions arrived almost simultaneously . . ., each different from one another, and both different from M. de Roberval's, though nevertheless in such a way that on looking at them all together it is not difficult to discover which came from the Author' (195) of the competition (i.e., Roberval): Pascal puts Fermat and Descartes in second place. In 1638, so the account continues (195), Roberval's discovery of the ratio of the area of a cycloid to that of its generating circle, together with Fermat's method of maxima and minima (which became the starting-point of Leibniz's differential calculus), were communicated to Galileo by Jean de Beaugrand:[28]

> He sent them both to Galileo, without mentioning their authors' names; admittedly, he did not exactly say they were by himself: but such was his choice of words that, unless you looked very closely and attentively, it might seem that it was only out of modesty that he had not put his own name to them; and, so as to disguise matters a little, he altered the earlier names *roulette* and *trochoid* to *cycloid* (195).

In December 1640 Beaugrand died, to be followed in January 1642 by Galileo:

> Torricelli followed in Galileo's footsteps and, having come into possession of all his papers, found amongst other things the solutions to the *roulette* alias cycloid problem, in M. de Beaugrand's handwriting, he apparently being their author; and, M. de Beaugrand being dead, Torricelli thought that sufficient time had elapsed for people to have forgotten about him – and decided to take advantage of the fact (195).

And so accusation is piled on accusation: that in 1644 Torricelli claimed that Mersenne's intuition of the cycloid problem was really Galileo's and that Galileo, not Roberval, had been the first to resolve it;[29] that both victims of Torricelli's plagiarism (Mersenne and Roberval) complained to him about it; that in letters to Mersenne Torricelli privately admitted his guilt without ever publicly abandoning his claims; and that Torricelli tried to save his face by solving the problem of the volume of the solid of revolution generated by the segment of the cycloid revolving around its axis; but failed again, and (in October 1647) died.

The outcry against Pascal was tremendous. Mathematicians, and not just Torricelli's friends, sprang up from England to Italy in the dead man's defence. As was pointed out by John Wallis,[30] Torricelli had a strong and genuine claim to originality; Pascal's attitude towards Torricelli, Joseph Bertrand remarks,[31] is marred by 'inexplicable prejudices'. Pascal had, in fact, rekindled the dying embers of a much earlier controversy between Torricelli and Roberval on the same subject, in which Roberval had accused the Italian of plagiarism immediately on the publication of his *Geometrical Works*. Here, as in the controversy with Descartes over the construction of the tangent to a cycloid, or in the still greater controversy involving Cavalieri and the method of indivisibles, Roberval's passion for secrecy did him a profound disservice. There seems, however, to be no room for doubt that Torricelli arrived at his resolution of the quadrature of the cycloid curve by independent methods (**VIII 184–94**).[32]

Why then did Pascal lend his name to what is probably such a travesty of the truth? And why does the *History of the Cycloid* contain no mention of Cavalieri, whose method of indivisibles[33] foreshadows both Roberval's and Pascal's? This second circumstance is no easier to explain than his alleged antipathy towards Torricelli, whilst the first no doubt arises from his close friendship with Roberval, whose word, on the vexed question of the quadrature of the cycloid, as on other matters, he was very ready to take on trust. But over and above the obvious willingness of one friend to spring to another's defence, there may have been a second reason for Pascal's somewhat ruthless denunciation of a scientist who was no longer alive to defend himself. For the controversies between Mersenne and Galileo, Roberval and Torricelli, lead on in the plan of the pamphlet to Pascal's own difficulties both with Wallis and La Loubère.

Not content with having abundantly demonstrated to La Loubère his own inadequacies as a mathematician, of which La Loubère himself was only too meekly aware, and as if mocking him unjustly for still seeking something he had already found were not enough, Pascal proceeds in the concluding pages of his *History of the Cycloid* (199–200) to set the poor man three further questions: all abstruse in the extreme, but which *he* had already solved. He gave La Loubère, and any others who might be interested, until 31 December 1658 to produce their answers (200).

La Loubère was no longer interested in trying his feebler strength against Pascal's genius. Having failed with the two major questions of the competition proper, he had no reason to continue. Nevertheless, his own limitations as a thinker did not lessen his acute sense of grievance as a human being. On 24 November the panel of judges had met, under Carcavi's chairmanship, to consider the two entries. Both were found wanting, and so the Anonymous challenger retained his prize money. On the following day a Report (208–10) was published, in measured and human terms, setting out the reasons for the judges' decisions. La Loubère, however, refused to accept these decisions, even claiming that he had been victimized. His stubborn feeling of outrage drew from Pascal one of the sharpest attacks he ever wrote against the erring waywardness of human character.

In this *Sequel to the History of the Cycloid* (211–16), published on 12 December 1658, La Loubère is denounced for his gross and almost total incompetence. He is mocked, too. 'Geometrical matters are so serious in themselves', Pascal writes as his opening sentence, 'that it is a good thing when an opportunity arises to make them slightly diverting': the tone is somewhat alarming. Then, in an exquisite blend of the lofty and the feline, La Loubère is excoriated. Not until Pascal's *Letters from A. Dettonville to M. de Carcavi* (224–303), written under the pseudonym of Amos Dettonville which is an anagram of Louis de Montalte, and published at the very end of December and beginning of January, were the solutions to all the cycloid problems – including the three additional ones – finally given. But La Loubère had still not overcome his burning resentment. The consummate beauty of Pascal's demonstrations may even have added to it. A few days after the appearance of the first section (224–46) of the *Letters to Carcavi* he wrote personally to La Loubère, on New Year's day 1659 (216), sending him a copy of it. La Loubère, it seems (216), replied in a pamphlet dated 9 January 1659, outlining conclusions suspiciously similar to Pascal's own. It would have been unlike Pascal not to voice this suspicion. In the *Footnote to the Sequel to the History of the Cycloid* (216–17), published on 20 January, La Loubère is accused of inaccuracies in his submission eleven days previously, especially in his calculations of the distance separating the axis from the centre of gravity of the semisolid of revolution generated by the upper segment of the cycloid rotating around its axis.

The priest rushed once more into print with a *Response* dated 15 February 1659 (**IX 169–70**), but nothing could now conceal his defeat by a man whose superiority over him was even greater in the controlled literary handling of words than in the *a priori* manipulation of mathematical symbols. Whether La Loubère's being a Jesuit contributed to Pascal's animosity can never be known. The fact remains that this Jesuit was no Étienne Noël, and that the competition which Roannez hoped would redound to the greater glory of the Christian religion ended ingloriously in the unedifying spectacle of two Christian believers at each other's throats. La Loubère is reported as expressing astonishment at the deterioration in their relations between September and December 1658 (**VIII 121**); but, for an explanation of this change of attitude, he need not have looked far. Pascal, out of his concern for truth, was shocked by the undeniable presence of errors in La Loubère's calculations. He could not help wondering how a man who could commit such enormities in September could possibly be capable both of accuracy and penetrating insight by the following January. La Loubère may or may not have plagiarized Pascal's demonstrations concerning the centres of gravity of the semisolids of a cycloid; it is understandable that he should have done so.

But Pascal likewise cannot be exempted from blame. The terms and conditions of his competition were a muddle. He, only four months before writing his *History of the Cycloid*, had been strangely ignorant of Roberval's work on the subject. There was no need to prolong and confuse the competition by setting additional quest-

Figure 10.1 The Fundamental Cycloid Problems

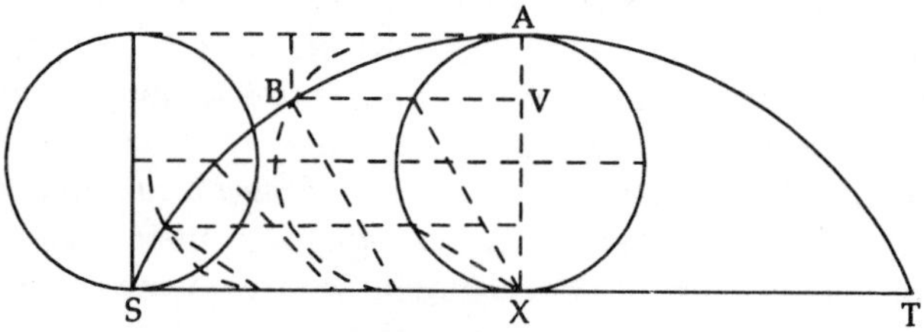

Given a cycloidal arc with base ST and axis AX, and given also the semicurvilinear surface ABV defined by the curve, the axis, and a semichord BV parallel to the base:

Pascal established
(1) the area of ABV and its centre of gravity;
(2) the volumes of the solids V_1 and V_2 generated by the revolution of ABV about AV and about BV, as well as their centres of gravity;
and (3) the centres of gravity of the semisolids obtained by intersecting V_1 and V_2 by midplanes.

From the *Second Circular Letter Concerning the Cycloid*
July 1658

ions. Nor was it any concern of his to intervene in the running of the competition, as he did in his two letters to La Loubère written as the deadline for submissions approached. Nevertheless, the importance of Pascal's geometrical discoveries made in such unusual circumstances during the last few months of 1658 is beyond dispute. The cycloid competition revealed him once again as an intellectual giant, though very far from being a giant of charity or humility. The *Letters from Dettonville to M. de Carcavi* are a Treatise on the Cycloid in all but name, immensely authoritative and fruitful in its influence upon the development of the integral calculus.[34] Their importance and originality lie not so much in their resolution of specific problems – many of which had been resolved, even before the start of the competition, by Roberval and another, during the competition, by Wren – as in the formulation of a general method. Inspired by Archimedes's method of determining the quadrature of the parabola by means of the equilibrium of the lever, Pascal's new and generalized approach to infinitesimal quantities enlarged the concept of the Archimedean lever to include unequal weights at intermediate distances from the fulcrum.

Pascal also goes well beyond Archimedes's work on parabolas and on the polygons used to determine the value of π, by concentrating instead on the segment of a cycloid and on the infinitesimally large number of infinitesimally small rectangles that are contained within it. Influenced by Archimedes, the Pascalian calculus became a model of rigorous simplicity. It concentrated on the interrelated concepts of the infinitely large and the infinitely small, perhaps also influenced in this respect by Kepler's *Supplement to the Stereometry of Archimedes*, which had introduced into geometry the concept of infinitely small magnitudes.

But the Pascalian infinitesmimal calculus fell short of Newton in that, however generalized its method in relationship to the cycloid, it dealt only with one particular curve without establishing the universal geometrical laws applicable to all curves, the areas of their segments, and the volumes of their solids of revolution. From Newton's generalized method of defining the areas and volumes of curvilinear figures came his celebrated method of fluxions (which, though invented in 1665–6, was not published, partially, until 1687 and, more fully, in 1704): this was the formula for determining fluctuating velocities by means of the calculation of variables in both space and time. In this matter again, a sturdy pillar supporting the complex edifice of modern mathematics, Pascal played no part.

His language in the *Letters to Carcavi* and their ancillary tracts was profoundly non-algebraic, yet, writes Nicolas Bourbaki,[35] 'without writing a single formula', he succeeding in making it 'so transparent, so precise that it can immediately be transcribed into formulæ'. Pascal, unlike Fermat and Descartes, had little time for algebra, merely acknowledging that it had its uses whenever demonstrations were 'difficult' (1428) – which he evidently did not feel the cycloid to be!

The *Letters to Carcavi* were by no means the sum total of his mathematical achievement published under that name around the turn of the year 1658–9. He was also the author of a *General Treatise on the Cycloid*, and of the *Treatises on Geometry*,[36] one of which was to open Leibniz's eyes to a profound mathematical truth. The latter, in a deleted postscript meant for Jakob Bernoulli in 1703,[37] acknowledges – at least to himself – his indebtedness to the 'Treatise on the Sines of Quarter-Circles', which he had come across, thirty years or so previously, during his time in Paris. Attributing his own use of the differential triangle to a figure he had happened to notice in that treatise, Leibniz states that upon seeing Pascal's figure he had a flash of insight: namely, that whereas the quadrature of the cycloid depends on the sum of the ordinates for infinitesimal intervals in the abscissas, the *differences* between the ordinates and the abscissas are what determines the tangent to a cycloidal curve.[38] This insight of genius was directly inspired by the 'Treatise on the Sines of Quarter-Circles',[39] yet it was something which the author of that treatise had failed to realize for himself. Pascal, Leibniz remarks, seemed scientifically 'blindfolded' in his failure to discern some of the applications of his work on cycloids. Blindfolded he may have been, but the truth hidden away beneath the unyielding exterior of the 'Treatise on the Sines of Quarter-Circles' was dazzling nevertheless. Pascal's influence upon Leibniz can only have been indirect, as he had been operating in the calculus of infinitesimals; yet the differential calculus, and the notation of the calculus used nowadays, would not perhaps have come into existence except for the *Treatises on Geometry*.

Pascal was also the author of various important mathematical letters to his friends. In an open letter of 10 December 1658 (313–27) to a mysterious correspondent disguised under the initials A.D.D.S. (and presumed to be Antoine Arnauld), the not quite so mysterious Dettonville demonstrated, against Hobbes, the equality of length of the parabolic curve and the Archimedean spiral: 'the only [demonstration] of its kind, no other one in the manner of the Ancients having yet appeared by way of a comparison of two lines of different natures' (314).[40] In an open letter to Sluse (328–34), written probably later in the same month (December 1658), Dettonville demonstrated certain properties of the so-called *escalier* (or *staircase*), the cylindrical triangle, and the solid of revolution generated by a spiral rotating around a cone; the *escalier* being a solid consisting of spiral steps of uniform height and surface area mounting within the arc of a circle and with a quarter of that circle as its base: Pascal established that the volume of this solid was equal to one quarter of the square of the base arc multiplied by the radius (333). Yet another open letter, this time addressed to Huygens and composed in January 1659 (335–40),[41] greatly enlarged the scope of the *Letters to Carcavi* by providing a method of determining the quadrature of *all* cycloids: not only the common ones, those to which Dettonville had confined his attention hitherto, but also prolate and curtate cycloid curves. A final letter to Carcavi (341–3) concerned the convex dimension of the parabolic conoid – which Pascal (reasserting his personal identity) hoped in due course to extend to the parabolic spheroid (1438): in vain, however, as his health did not permit it.

Now indeed, with this letter to Carcavi, the end of the mathematical research had come. Even after his second conversion Pascal's secular intellectual work had been continuous if intermittent. In the autumn and winter of 1656–7, two years or more after the night of fire, he had been discussing with Fermat and Carcavi peculiarly difficult aspects of the probability calculus. His very substantial achievements in the geometry of cycloids were not some weird flash of inspiration (all over within two hours, like the mystical vision) during a night of acute physical torment. The very length and density of the *Letters from Dettonville to Carcavi* and their appendices are evidence of hard and sustained effort over many weeks. Although a bout of toothache may have brought Pascal back to the study of geometry, the work on cycloid curves was far from being his first return to mathematics since his mystical conversion. It was, however, to be his last.

11

Conclusion

Basing his theories on quantitative experiment, Pascal was to that extent a greater natural scientist than Descartes. (Newton, towards the end of the seventeenth century, was perceived by many to have refuted the vortex hypothesis by resorting to mathematics.[1]) A spur to the younger Frenchman's achievements in physics was that he responded vigorously to the stimulus of his scientific colleagues, taking part with them in the informal scientific discussions chaired by Mersenne and Le Pailleur, arguing ferociously with Noël and setting challenges to Fermat, Wallis, Sluse and Wren. Towards mathematics he was as pragmatic in his approach as he was towards physics. His findings in projective geometry were inspired by Desargues. The work on the probability calculus began with his response to Méré's enquiries about the number of throws of two dice probably required to come up with a double six and how best the stakes should be divided in the event of the interruption of a game of dice; it was fulfilled in the five letters exchanged between Fermat and himself. This way of reacting *ad hoc* to external stimuli has been characteristic of the achievement of many scientists, from Newton's intuition of gravitational force on supposedly seeing an apple fall in his Lincolnshire orchard to Fleming's discovery of penicillin on seeing a chance growth of mould in a culture of staphylococci. But such a reaction to external stimuli was peculiarly characteristic of Pascal in that without them, and above all without the challenge to tower over other men, he did not seem keen to embark on scientific work at all.

Pascal's preoccupation with number was turned to brilliant advantage when he invented his mechanical calculator; as so often in his career, this discovery responded to a particular practical need. From his Puy-de-Dôme experiment in September 1648, which once again was based on careful observation and quantitative experiment, sound practical benefits arose: the invention of the barometer, the creation of the science of meteorology, a method of calculating altitudes, and (in the longer run) the invention of hydraulic brakes, pumps, turbines, lifts, cranes, the syringe and modern anæsthesia.

Pascal's discoveries in physics, though perhaps rather less remarkable than the great things he was yet to achieve in mathematics, have been extraordinarily beneficial in their impact upon the history of the world.

In mathematics he is notable for his work on arithmetic, probability theory and combinatorial analysis. The research into the two latter subjects has been of inestimable importance during the last three centuries. Combinatorial analysis, which became Pascal's sphere more than it was Fermat's,[2] has deeply influenced actuarial work, cybernetics, psephology and operational research. The calculus of probability, Pascal's and Fermat's joint field of research, has had a far-reaching impact upon more recent physics and philosophy. It has shaped quantum mechanics and, by releasing the physical world from the requirement of absolute certainty, has paved the way for the discoveries of mathematical physicists such as Clerk Maxwell. Statistics, economics, games theory and, more generally, all spheres in which the concept of calculable risk is involved are hugely indebted to *The Probability Calculus*. Pascal's arithmetical work, on the other hand (which was mostly on interests he shared with Fermat, such as prime numbers and magic squares), may have been a sort of 'art for art's sake' of mathematics in that it was, and could only have been, of limited practical benefit to mankind. Finally, Pascal's work on the quadrature of the cycloid curve created the infinitesimal calculus, although (failing to establish the geometrical laws applicable to all curves) it did not achieve the generalized formulation which made Newton's integral calculus possible.

The fact that Pascal did not discover a method of fluxions, or something like it, is significant in a wider context. Though he may have believed in some sort of unitary philosophical system embracing the whole of the scientist's and the mathematician's world, he did not actively seek a system whereby the laws relating to *all* natural phenomena, or *all* geometrical figures, could be established. As a scientist, he chose rather to infer from directly observed evidence; this led him, for example, to doubt Descartes's hypothesis of vortices. The transition from the old intellectual order to the new is well illustrated not only in the matter of vortices but also, and first and foremost, in the dispute with Étienne Noël. Utterly at variance with this entrenched Aristotelian, Pascal, as an inaugurator of the modern scientific tradition, was close in spirit to his forerunners Stevin, Torricelli, and above all to Galileo who in

1633 had been so fiercely reprimanded by a Church which, in matters of physics, was still the custodian of Aristotelianism.

On the other hand, and also because of this, Pascal's scientific work is equally remarkable for its modest aims and its comparative freedom from metaphysical presupposition. Had he attempted to provide a complete world-picture, he would have had to draw upon many more presuppositions than was required for the solution of individual problems successively. But to offer, or to draw deductions from, such a world-picture was well beyond his sphere of activity as a scientist. His interpretations of his experiments do not depend for their cogency upon the validity of a particular world-view. From this point of view his writings contrast strongly not only with Noël's utterances but even with many of Descartes's.[3] There is, as a matter of fact, only one point at which Pascal makes a cosmological presupposition, and it is only the merest shadow of one. This occurs when, in *New Experiments Concerning Vacuums*, he seeks to prove the existence of vacuums from the tendency of the air to rarefy and condense (363): for such an argument holds good only in respect of an atomic world. Otherwise, however,[4] his hypothesizing is strictly limited to what he has observed.

In his research in physics he subordinated deductive or *a priori* reasoning to the inductive or *a posteriori* method of inference from observation. Whilst not ruling out the value of intuition when conjecturing hypotheses in the realm of physics, he himself preferred to test by means of experiments the conjectures which had been advanced by Stevin, Mersenne, Galileo, Torricelli and others. In the sphere of inference from recorded observation he would not, however, admit the value of propositions that are merely probable: something astonishing in a man famous for his writings on mathematical probability. This fact, indeed, marks him out as a man of the older scientific covenant; the newer way of dealing with scientific hypotheses begins with Leibniz, whose views on finite deductions and the probability calculations of inductive logic allow for hypotheses to be tested rather than (as in Pascal's case) verified.[5]

Pascal's technique of argument in the *Thoughts* recognizes (to a far greater extent than Descartes) that not only will the mind refuse its assent to the truths most worth knowing unless the heart has been touched but also that some truths of the mind cannot be perceived, still less retained, except through the conjoint promptings of mind and heart. It cannot therefore be encompassed within the conventional (Cartesian) framework of tightly knit syllogism and necessary consequence. His methodology is not, therefore, that of the traditional philosopher; and in this respect his philosophical method is profoundly original. He has no wish whatsoever to construct a formulaic metaphysical system.

He even professes his indifference towards the truth, or falsehood, of Copernican astronomy (164*):

> I think it is right and proper that one should not look deeply into Copernicus's opinion.[6] However:
> It affects our whole life that we should know whether the soul is mortal or immortal.

Likewise, Descartes – 'useless and uncertain' (887*) – has, he claims, '[gone] into science too deeply' (553*).[7] These statements have provoked much adverse comment, not to say blank amazement. The Copernicus remark echoes Montaigne,[8] however, and may belong to approximately that time in Pascal's life when he rated geometricians no more highly than honest artisans, teasingly proclaiming that whilst geometry was a good means of testing one's mental powers, it was not a good means of employing them (522).[9] It is also true[10] (as he had realized, and actually argued, in his controversy with Noël: 375[11]) that the Copernican system of celestial mechanics was merely one of three possibilities, the others being Ptolemy's and Tycho Brahe's, none of which, in the then state of scientific knowledge, could be verified at the expense of the others and thus elevated to the status of a generally acceptable theory.

The fact nevertheless remains that the absence of a general method for which Carré has criticized Pascal in the realm of religious persuasion[12] is also conspicuous in his scientific writings. These words about Copernicus would be strange indeed from any scientist concerned to build up a complete explanation of the natural phenomena of the universe. For all his great experimental gifts, Pascal was not possessed of an insatiable scientific curiosity.

Within the framework of European science Pascal and Descartes had much in common. Both believed that at the heart of science, and in no way distinct from it, was mathematics. Neither, then, was first and foremost a natural scientist. Both believed (but Descartes more enthusiastically than Pascal) that not only arithmetic, geometry and algebra but also astronomy, mechanics, optics and the whole of what today would be called the natural sciences constituted, or formed part of, a unitary system of principles and laws. Both believed that these principles and laws can be expressed in quantitative terms. Both tended to reject specific explanations of individual phenomena that were unrelated to the general whole. Most importantly of all, since both believed in a system of principles and laws that linked mathematics to the natural sciences, so both believed in deductions from first principles: both, in other words (but Descartes more enthusiastically than Pascal), applied the 'geometric' method to their study of the physical world.

In pure geometry Descartes had in 1647, and always retained,[13] the pre-eminence over his younger contemporary. This was the field of mathematical study which at that time had by far the longest and most distinguished lineage. It extended back to Archimedes, Apollonius of Perga, Euclid and Pythagoras, whereas algebra was a more recent importation from the Arabic world, a science with only just over four centuries of European history. But it was Descartes's self-enclosed, self-contained deductive system of co-ordinate geometry which set the entire subject on its more algebraic path, whereas Pascal, who lacked the algebraic mind, remained, in his devotion to projective geometry, somewhat rooted in the past. And it was, of course, for this reason that the really large strides in the calculus were taken by Newton and Leibniz, the way forward to the integral calculus and the differential calculus respectively having been suggested to them by Descartes.

For both Pascal and Descartes the term 'geometric method' meant the habit of deducing *a priori*, or from first principles. Very similar to Spinoza's method in his slightly later work *Ethics*, it was close to what today would be regarded as an algebraic or syllogistic method of thinking. The 'geometric' method was not directly concerned with properties of space. Nor was it a method of inference.

Centred upon mathematics, the Aristotelian universe had been a close-knit self-authenticating system of truth, a system the parts of which were linked by demonstration. The neo-Aristotelian system was still the Jesuits' intellectual province; it presupposed that knowledge or *scientia* was arrived at by demonstration from first principles. Descartes was as much a product of the Society of Jesus as was Étienne Noël. Within the neo-Aristotelian system he believed that truths existed independently of proof, whereas for Leibniz proof constituted truths. The process of deduction necessitated, in Descartes's view,[14] that initial propositions should be intuited: the further steps, of working out the laws governing the behaviour of the physical world, were consequent upon this initial intuition. Whereas intuitionism underpinned Descartes's philosophy, Leibniz was the first formalist[15] – and so is regarded as the creator of symbolic logic. It is clear, however, that by the year 1638 Descartes had also become aware of the process of explaining observed effects by means of the postulation of causes.[16]

The vacuum controversy, whilst illustrating the intuitionist aspect of this argument perfectly, also underlines another feature of Pascal's controversial encounters with Descartes. Had Descartes accepted the concept of vacuums (which, at any rate in September 1647, he was unwilling to do), it would have meant that God, through a long chain of consequential events, had caused vacuums to be created. Pascal saw no difficulty in this. His work on vacuums embodied the outcome of objective experimentation. Being the very reverse of a neo-Aristotelian, he sees no contradiction between belief in vacuums and belief in God.

In the highly contentious matter of vacuums and atmospheric pressure debated by Noël and the Pascals (father and son) in four notable letters, the much younger layman conducted experiments, inspired by Stevin, Mersenne, Galileo and Torricelli, which had the strictly limited objective of testing by observation and inference. This led him to assert the existence of vacuums, and even to deny the concept of Nature's limited abhorrence of the void. It also led him to confirm the role of the external air column in his experiments (and thus to accept, as by far the most valid hypothesis, the principle of atmospheric pressure) and to state the principle – Pascal's Principle – that hydrostatic pressure is the same in all directions about a point in a fluid. The Descartes of *The Principles of Philosophy*, on the other hand (though not the Descartes of letters to the mathematician Jean-Baptiste Morin and perhaps also Mersenne),[17] denied that vacuums existed and to that extent his mind was not open to the evidence of observation and induction.

To many readers of Pascal from 1670,[18] and particularly from 1740,[19] onwards it must indeed have seemed a paradox that Descartes, holding the weaker notion of God, denied that vacuums could exist; whereas Pascal, who asserted the existence of the void, was imbued with much the stronger notion of a personal Divinity. *Thought* 463* comments on this paradox:

> It is a remarkable thing that no canonical author has used an argument from nature in order to prove the existence of God . . .

> David,[20] Solomon[21] etc never said: 'The void does not exist . . . Therefore there is a God.'

At a time when the verbal inerrancy of the Scriptures was a cornerstone of religious belief, this, to Pascal, was an important point. Directly inspired by God, the Biblical authors do not pronounce judgment on the void; fortunately for themselves, they do not make the non-existence of vacuums a cornerstone of religious belief. The same preoccupation underlies Pascal's notable quip (1001*) that all Descartes's God ever had to do was just to snap His fingers and so set the universe in motion. There is no ultimate purpose in the Cartesian system. Unlike Pascal's universe and that of the medieval schoolmen (St Thomas Aquinas, Albertus Magnus, St Bonaventura, Alexander of Hales and Duns Scotus) who had been inspired by Aristotle, Descartes's universe is presided over by a God Who is the reverse of teleological, being the Prime Mover of a mechanically functioning universe which is material rather than spiritual and not subject to His providential control: a First Cause primarily important to man as the ultimate Guarantor of the human thinking which was also the clearest evidence of His own existence.[22] According to the schoolmen, on the other hand, the divine purpose of the world was that, through man, that world might return to God Himself; whereas Pascal's God resembles Newton's in that He is the supreme embodiment of voluntarism, actively intervening in human history through the exercise of His personal will. Pascal did not believe in the austerely rational God of the Cartesian system, nor in chill rational processes of attaining a knowledge of Him. Although both he and Descartes acknowledged the huge importance of reason in human life and affairs, the younger man's caution as to the exercise of human reason in a contingent world may have led him, even at this early stage of his philosophical thinking, to doubt the value of any knowledge of God the Supreme Artificer which was not mediated through the Person of Jesus Christ.

In two important passages Pascal refers to the 'geometric' and the non-'geometric' methods: in *On the Geometrical Mind* (575–92) which, with its counterpart *On the Art of Persuasion* (592–602), was probably written early in 1657; and in the celebrated *Thought* 512* concerning *l'esprit de géométrie* and *l'esprit de finesse*, which probably dates from the following year. From these passages it emerges that Pascal is less adamant than either Leibniz or Spinoza in his commitment to the ideal of absolutely certain knowledge: knowledge that flows from some self-evident postulate, or postulates, with logical precision. For him, however (much more so than for Descartes), the scope of the deductive method is limited. He attaches great importance to experimentation in the belief that the truths of physics are such that they cannot be established deductively even if that is one method of subsequently presenting them.

The *esprit de géométrie* is, of course, a faculty of mind which few people possess in any high degree. Few can understand Cardano's Rule, Mersenne Numbers, Desargues's Theorem or Cavalieri's Principle. And because of this, says Pascal, some human beings (such as are interested in these matters at all) prefer to concentrate on the study of human nature. Yet even here, he claims, their analysis is woefully inadequate. *Thought* 687* is central to his thinking in this respect:

> I spent a long time studying the abstract sciences and had been put off by the fact that they have so little to tell us. When I began the study of man, I realized that these abstract sciences were not mankind's proper concern and that, by analysing them, I was wandering further away from my human condition than were other people by overlooking them. I forgave others for knowing little about them, but at least I thought I would meet with many fellow-students of mankind and that this was the thing really worth studying and man's proper concern. I have been mistaken. There are even fewer people studying this subject than there are studying geometry. It is only because they do not know how to study the latter that they turn to the other. But is it not true to say that even this is not the knowledge that a man must have, and that it is better for him to know nothing about himself if he is to be happy?

Though few, according to Pascal, can understand, fewer still can comprehend. The study of human nature is no simple matter: it is, he argues, even harder to obtain an adequate understanding of this than it is to obtain an adequate understanding of mathematics. And, if a man is to be happy, it is best for him to know nothing about human nature in the abstract and certainly to avoid introspection: this echoes the notion of *divertissement*. Pascal evidently considers that he has attained a proper understanding of man – that his competence in the sphere of *finesse* is as great as it is in that of *géométrie*. And yet, if we view his enquiry into human nature by any of the objective criteria whereby the truth of a discipline should and must be judged, there are no benchmarks by which the accuracy of Pascal's analysis of human nature can either easily or definitively be established. In this respect he is reminiscent of Rousseau, whose *Confessions* were intended as a contribution to the understanding of human nature. Human nature or a human nature? The competition, in 1658, concerning the surface area, volumes, and centres of gravity of a cycloid's segments and of its solids of revolution could be resolved by reference to Pierre de Carcavi, the chairman of the panel of judges; for it involved matters of *géométrie*, not *finesse*. And no one denies that Pascal has made major contributions to the study of the cycloid curve, probability theory and the science of atmospheric pressure (less so to conic sections). These contributions are a matter of record; their only contentious aspect is to decide how important a place in the history of mathematics or physics they actually confer upon Pascal. But is his study of human nature even-handed? Contrariwise to the matter of Pascal's scientific status, no one will deny his outstanding and distinctive gifts as a weaver of words portraying the human condition as he saw it:

> Quelle chimère est-ce donc que l'homme? quelle nouveauté, quel monstre, quel chaos, quel sujet de contradictions, quel prodige? Juge de toutes choses, imbécile ver de terre, dépositaire du vrai, cloaque d'incertitude et d'erreur, gloire et rebut de l'univers.[23]

> Le dernier acte est sanglant, quelque belle que soit la comédie en tout le reste. On jette enfin de la terre sur la tête et en voilà pour jamais.[24]
>
> Le silence éternel de ces espaces infinis m'effraie.[25]

But, in this sphere of *finesse* which by his own admission is exceedingly difficult to comprehend, is Pascal's analysis of human nature an accurate one? On its accuracy depends, it would seem, the case which he is making for the Christian religion. Yet there is no consensus as to the accuracy of his findings: Pascal, as was seen in Chapter I, has appeared at various times and on various occasions as a 'remarkable, or rather matchless, intellect', 'a sublime misanthropist', 'proud', 'arrogant', a 'Christian hero', corrupting, neurotic, of 'pessimistic natural temperament', 'heaven-born', 'extremely pernicious', 'mad', 'domineering' and a 'skinflint . . . bristling with hatred'.

Is his human nature, which he analyses introspectively, a human nature or the paradigm of human nature generally? In this respect it would seem that Pascal, who had never really sought it in the abstract sciences, is too keen on the unifying solution: one human nature not merely being seen as part of the whole but as symbolic of the totality of human nature. He was a sick man, never knowing a painless day from the age of eighteen: can his perceptions of living be representative of the whole? Or is he not like Leopardi, so ill that his physical infirmities colour, or discolour, his impressions of the joy of living? Does not his neurosis (in Daudet's phrase)[26] cause him to look on life with a cold eye? Is it not this neurosis which leads him to say that no man can sit happily and quietly in his own room (136*), and, more importantly still, to claim that herein lies the misfortune of the human lot?

> Man finds nothing so intolerable as to be in a state of complete rest, without passions, without occupation, without diversion and without effort.
> He then faces up to his nothingness, loneliness, inadequacy, dependence, helplessness, emptiness.
> And immediately there well up within him boredom, gloom, depression, vexation, resentment and despair (622*).

Pascal's outlook is that of the countless funeral monuments of the sixteenth, seventeenth and eighteenth centuries where, above or beside the recumbent effigies of the departed, a tangible reminder of man's inescapable mortality is to be seen: *memento mori*. However, availing himself of the medium of words rather than marble or alabaster, he far outstrips the message of church monuments, likening life on earth to a condemned cell in which all sit waiting for the sentence of the court or the executioner's possible arrival (163*, 164*): you will be executed, so Pascal's message runs, unless you reform your life, in which case the sentence can be averted.

Nevertheless, it would seem that many human beings can sit quietly and happily in their own rooms. Without invoking our own human acquaintance, we need think only of examples taken from literature, of Disraeli's Contarini Fleming claiming that 'the sense of existence is the greatest happiness',[27] Rousseau arguing likewise in his letter to Voltaire about the Lisbon earthquake,[28] or Meursault, in Camus's *The Outsider*, sitting quietly and happily in his own room or lying contentedly on the beach. But was the reality happy, Meursault and Contarini Fleming being after all mere figments of their creators' imagination, literary characters devoid of existential existence? Or is the practice of literature, like that of painting (40*), simply a form of *divertissement*?

There can be no question but that death overhangs the whole of the *Thoughts*, just as it overhangs the whole of our lives. It is one of only two omnipresent human realities, birth being the other. But whereas the *Thoughts* are overhung with the notion of death as an existential challenge (434*), this same notion of challenge does not necessarily overhang the way in which all people envisage their own lives. And this again is the result of *divertissement*: it is, says Pascal, the consequence of the fact (if fact it is) that many people will engage in any number of frivolous, or allegedly frivolous, occupations – gambling, hunting, feminine company, diplomacy, warfare (136*), theatregoing, paying social calls (628*)[29] – in order to avoid facing up to the truth about their own human nature and the grim omnipresence of death.

This is the sense in which human life, as envisaged by Pascal, must be considered a tragedy, for (according to him) all, in an act of anagnorisis, must face up to the truth about death, human life and their own selves. But what is this truth about death? Is there not merely the (self-evident) truth that death is a mystery which may involve certain consequences? And thus we are led to the Wager, the second manifestation of the way in which death overhangs the whole of Pascal's *Thoughts*. Was it because of this, and because so many people blithely disregard their futurity, that Pascal claimed so few people were proficient in the study of human nature?

The Wager, couched in the language of mathematical analysis, is no doubt intended thereby to possess an unchallengeable intellectual authority. Pascal's study of the *moral* sciences (Pascal *moraliste*: the student of human nature) is profoundly conditioned by his study of the *abstract* sciences of mathematics and physics. That unchallengeable intellectual authority which he knew was attached to the discovery of properties of the cycloid curve could, he felt, be transferred to the *moral* sciences without let or hindrance: the Duc de Roannez may have encouraged him to think this (40). It might, Pascal believed, be possible to *prove* one aspect of the truth of Christianity with exactly as much certainty as it can be proved that, if the area of a circle is equal to πr^2, then the circumference of a circle equals $2\pi r$. And this aspect of the truth of the Christian religion which could be proved in the manner of the abstract sciences was, he surmised, the Wager. Not even Pascal the abstract scientist ever asserted that it was possible to prove anything else about the Christian religion; but the Wager, he believed, might be provable beyond any doubt. For once the rigorous precision of *géométrie*, which few people can understand, could be brought to bear upon the much finer questions (comprehended by still fewer people!) of human mortality. This is the most exciting feature of the Wager and, in a way, the most remarkable thing about the whole of Pascal's stupendous achievement.

Certain obvious limitations of the Wager do not seem to have struck Pascal. What, in the Hereafter (if it exists), are we hoping for? 'An infinity of infinitely happy life'? Or could it be the abode of demons, or of an unjust God Who was not the 'God of rewards and punishments' that was Voltaire's ideal? It is by no means certain that a future life would be any improvement upon the present one: indeed, life after death could well be something less pleasant than the one on earth.[30] There can be no certainty then, within the terms of the Wager, that that for which we are wagering is 'an infinity of infinitely happy life'. To assume that life, if it is infinite, will be infinitely happy is to subscribe to the belief, widespread in the seventeenth century, that this is a rational world; comparatively few in that century expressed dissent from such a belief although there is evidence of it in, for example, Racine's tragedies.

Secondly, what is the meaning of 'wager'? A wager, as commonly understood, is the voluntary (and perhaps finely calculated) risking of a finite quantity in the hope of gaining more of that, or a similar, finite quantity. Instantly, therefore, we see that Pascal's Wager differs from the conventional model. It does so partly by virtue of the fact that the wagerer is being made to wager: surely this invalidates the concept of calculated risk? For what, after all, is the wagerer risking? Nothing, according to Pascal; but perhaps everything, according to the reluctant wagerer! And why is it more necessary that I should wager in the circumstances postulated by Pascal than that I should gamble on horses, which (let us say) I do not do?

Pascal's Wager also differs from the conventional model of the gambling situation in that like is not being compared with like, for in the conventional wager the only difference between the stake and the hoped-for gain lies in the order of magnitude (£1 being staked, for instance, in the hope of winning £10). According to the conventional model of the wager the hoped-for gain can, in other words, be clearly visualized: it is more of the same thing. According to Pascal's argument the Hereafter will, on the other hand, be different from the present, though he certainly does not make due allowance for the possibility that any future life would be less pleasant than the life we live now. To some people indeed, regardless of any doctrine of Purgatory, the prospect of living an eternal life somehow retaining the personal characteristics of life upon earth is a thing unbearable to contemplate, let alone to endure.

Furthermore, according to the commonly understood notion of the wager, the stake is but a small portion of what makes up the gambler's whole life, whereas, according to Pascal's Wager, the stake is the whole life – or rather, in his view, nothing at all, because (he insists, loudly and always) the ordinary human lifestyle has no value. It is also true in a mathematical sense that this is a comparison of disparates, in that the stake is (said to be) finite whereas the hoped-for gain is (perhaps) infinite.

There are indeed other objections and qualifications to be made about Pascal's Wager, such as, first and foremost, the charge that there is nothing in Pascal's line of reasoning to prevent the unbeliever from wagering only at the very last moment, just as a man wagering on a horse can decide when he will place his bet. Another objection is that it is a mercenary argument (close, therefore, to the mechanical view of religion for which Pascal berates the Jesuits) yet also – echoing, oddly enough, the Jesuits' charge against the *Provincial Letters* (779; *163*) – an irreverent argument in that it debases sacred things to the level of ridicule. Furthermore, it is open (though not perhaps vulnerable) to the criticism, first voiced by Condorcet,[31] that the wagerer would have to practise not merely the Christian religion but all religions that had not been shown to be false on other grounds. Yet another objection is that for some people it is quite literally impossible to change their hearts, even if by the light of reason they can change their minds: this is a difficulty which Pascal tries to counter by means of his concept of automatism. Again, granted that the whole tendency of the argument of the Wager is to persuade unbelievers to maximize gain rather than to minimize loss, it is nevertheless surprising that there should be no reference to Hell – the more so as Hell is invoked elsewhere (152*). Finally, however, Pascal's Wager raises the difficult question of freewill which to many people is the most impenetrable yet all-important question facing a human being. Some are as troubled by it as they are by the question of death. For many, indeed, during the late nineteenth and twentieth centuries the problem of freewill may well have taken on the intellectual precedence formerly held by the question of God's existence. In this connection the Wager offers a startling paradox: it is said that 'we *must* gamble' and that we are compelled to do so because we have 'embarked' upon our voyage through life; nevertheless, despite this element of compulsion, it is also strongly implied in the *Thoughts* that human beings are invested with the gift of freewill.

The indictment of so-called shortcomings in the Wager argument – 'a thought-provoking parable', according to Oystein Ore,[32] but without 'convincing mathematical weight' – appears to be endless, yet there is perhaps more sense in this argument than Pascal is usually given credit for. Although it is true that like is not being compared with like in the Wager argument, is there not also a similar comparison of disparates when £1 is staked on a horse in the hope of its winning £10? For is there not as little resemblance between the great racing victory and the solitary wagering of £1 as there is between life on this earth – however painful, constrained and seemingly futile – and the supposed glories of a life to come? The only connecting link between the two horseracing situations is the cash nexus; there is no such nexus in the Wager argument. Pascal's Wager may indeed be a very unusual example of the genre, yet it is a gamble that is being undertaken in very unusual circumstances. Underlying the Wager argument is a strong moral tone, which is why, in Pascal's view, 'we are compelled to gamble': we 'have to' do so because we are 'embarked' on a life of moral imperatives.

As previously with the mechanical calculator and the research into probability theory, and as he was later to do with the prize questions on the cycloid, Pascal harnesses his immense mathematical and scientific talent to a practical purpose: the probability calculus is applied to the soul's salvation! It is, of course, yet another paradoxical feature of the Wager argument that, according to him, nothing is being staked in order to win something (and something infinite). Here again, if we are to take Pascal's argument at its face level, the element of calculated risk which is at the heart of wagering seems to be absent. Apparently, therefore, Pascal is inviting man to undertake two (interrelated) gambles: first, that human life is worth nothing; and secondly, that by staking that nothing, a glittering prize may be won. The whole of the Wager argument is founded upon his own personal conviction that human existence is a poor insubstantial thing – not that this conviction is necessarily shared by others. And so, from the standpoint of his own negative view of terrestrial life, he seeks in every possible way to undermine hedonism; and this he does regardless of the possibility that the aching void in the heart of man (from which he sees *divertissement* as the means of escape) may just be the ardent yearning to achieve fulfilment of one's human, terrestrial life in whatever manner one chooses.

Indeed, within only about ten years of Pascal's death Pierre Nicole had voiced strong objection to the *Thought* (136*) on *divertissement*, claiming that the actual refusal to think – whether about oneself or about other matters – is the real cause of the painful sensation of living and that Pascal has confused two basic human emotions. 'Boredom', he argues, 'seeks *divertissement* whereas sadness shuns it'.[33] He maintains that this painful sensation of living (or *spleen* as it was known to Baudelaire, and *Weltzschmerz* as it was known to Jean Paul Richter and other German Romantics) can best be overcome by mental activity, providing that such activity is 'cheerful and agreeable'. Disregarding the fact that not all people find thinking an agreeable activity, Nicole makes no allowance for the human desire for escapism, nor for the possibility that there may be a deep well of dissatisfaction within the human heart.

Pascal, on the other hand, believing more strongly than Nicole in the nothingness of human life, sees the demolition of others' hedonism as his goal; nowhere, however, does he effectively deal with the converse possibility, that of nihilism. There is nothing in the *Thoughts* to lift man out of the dejected belief that human existence is meaningless to the point of absurdity, except perhaps that the Wager argument provides an escape from the view that human life has no self-evident meaning. But pessimism of this kind was basically foreign to Pascal's own way of thinking. Here was a man whose fundamental impulse was to invest his every action with meaning, even imparting extraordinary meaning to the solipsistic processes of mathematics.

The final paradoxical dimension of the *Thoughts* involves Pascal's notion of freewill. Does he believe in freewill? And how does he reconcile the idea of freewill with that of being *compelled to wager*? These questions would not appear so difficult if Pascal had not also to some extent defended the tenets of Jansenism.

We are all of us aware within ourselves of the strong feeling, mistaken or otherwise, that we are free agents.[34] Such is the foundation of democracy, freedom of speech, religious toleration; such too is the basis of the punishments meted out by courts of law. It may indeed seem to us at times as if there is a superabundance of possible choices, offering themselves at every turning in life's path: choices so various that the only philosophy to encompass them all is tychism. Descartes, whilst propounding a doctrine of necessity, also believed in human freedom;[35] Spinoza likewise.[36] Whether or not Pascal accepts any form of tychism it is frankly impossible to say; what is clear enough, however, is that he does not accept any doctrine of rigorous necessity so harsh as to preclude the spur to human action and self-improvement.

Expressed in another way: God (as Pascal sees it) may have foreordained the Elect, He may have decided 'before the foundation of the world'[37] to whom He would grant the gift of His grace; He may well have decided who shall be damned. To say that all this occurs thanks to divine foreknowledge would perhaps be an exaggeration from any point of view. For why should God foreordain the damnation of a soul? And if He foresees that damnation, is He not also foreordaining it? Even tychism, it would seem, is a more acceptable – and more intellectually creditable – proposition than this! It is in any event very hard to discern how Pascal, with the supreme emphasis he lays upon the wagering element of conversion, can so readily condemn semi-Pelagianism or even Pelagianism itself, except that the former is inextricably associated with the Society of Jesus. If he does not believe in freewill, what then would be the point of trying to convert unbelievers to the Christian faith? What, above all, would be the point of the Wager argument? Although both St Augustine[38] and St Thomas Aquinas appeared to set little store by human freewill (the latter claiming, for example, that 'from all eternity some are preordained and directed to Heaven; . . . others will not be given grace'),[39] the Wager argument would seem to be on much firmer ground when viewed from the standpoint of human freewill than from that of the possible nature of the Hereafter.

From the more recent point of view of the natural sciences and scientific determinism – that, for example, of Peter Atkins, Stephen Hawking, Richard Dawkins or Paul Davies – it is, of course, true that everything (except perhaps the cosmos) has a cause. Faced with a choice between A and B, a man may choose A: there must, however, be a cause, or a reason, for his choosing A. (Pascal's views on this matter do not engage with Hume's,[40] nor, it would seem, do the views of late twentieth-century scientists.) But in what sense can it be true to say that because everything has a cause, everything is predetermined? Tracing back, by a causal chain, the choice between A and B to its earliest causes, would it be possible by means of a sufficient number of calculations to predict with total accuracy whether a man will choose A or B? Pascal, committing himself to writing the *Thoughts*, thereby expressed his conviction that that work might cause people to believe in Christianity: it too, therefore, is a link in the long ætiological chain – not that the outcome of this long chain could be foretold *ab initio.*

First and foremost, however, the process of writing the *Thoughts* may justifiably be regarded as a facet of the divine foreknowledge and to that extent there is no contradiction between the freedom to convert and God's foreknowledge of the Elect and the damned.

In sum, there is very much to be said in favour of Pascal's overall concept of the *Thoughts*, including the argument of the Wager. 'The starting point for the religious attitude', T. E. Hulme has written,[41] '[is] always the kind of discussion you find in Pascal'. This starting-point, for most readers of Pascal and even for many who scarcely know of his writings, is the *Thought* on *divertissement* in association with the Wager argument. More clearly than anyone before him, Pascal sees the uselessness of what Newman was later to term 'paper logic':[42] again in Newman's phrase (as applied, however, to his own justification of Christianity), Pascal's technique is to build an edifice of 'powerful and concurrent'[43] reasons whereby, though no reason is in itself all-conclusive, the combined effect of all is to speak powerfully to both mind and heart. Perhaps the main such 'reason' is the Wager. Another 'powerful and concurrent' reason is the *Thought* (149*) of the Hidden God borrowed from Isaiah XLV 15. Another is the *Thought* (310*, 322*, 457*) of the 'deceitful Apostles' according to which it is inconceivable that Peter, James, John and their companions would have persisted in telling the same story, had that story been false. Yet another 'powerful and concurrent' reason is the *Thought* (308*) of the three orders. Another is the *Thought* (821*) on automatism, perhaps the strongest of all 'concurrent' reasons in that it is directly in keeping with New Testament teaching: 'as newborn babes, desire the sincere milk of the word, that ye may grow thereby'.[44] But over and above all such 'concurrent' reasons is the authority of the New Testament itself.

The fact that the Scriptural message has been handed down from the Apostles, through so many centuries and within the interpretive tradition of Church, Councils and Papacy, is for Pascal not only a guarantee of the Apostolic succession and Holy Orders but also of the absolute truthfulness of the Church's teaching. Voltaire, a century or so after Pascal, attacks (in, for example, his *Philosophical Dictionary*) what he considers to be the mumbo-jumbo of religious doctrine; Montesquieu, in *Persian Letters*, had similarly objected to the doctrines of bread becoming body, wine becoming blood, and Three being One;[45] Pascal sees the matter quite differently, however, because (in his view) not only is the Church the conduit and repository of doctrine but, in its sacred writings, in the fulfilment of their prophecies and in the witness of their miracles, it underpins the leap of faith of the Wager. He has to acknowledge that the so-called authority of the New Testament will probably mean nothing to an unbeliever; but once the first step in commitment has been taken, the Scriptural tradition (*deus absconditus* and all) will, he hopes, reinforce that commitment with incontrovertible clarity. Thus, with Tertullian, it will no longer be necessary to say: 'certum est *quia* impossibile est',[46] for the Scriptures will reveal the very opposite of the impossible or the absurd. The 'Word within the world unable to speak a word'[47] in the manger at Bethlehem, far from being an absurdity, becomes thereby the focus of God's mysterious self-revelation.

Pascal, as a matter of fact, carries little doctrinal ballast as Christian apologists go. He has little to say about the Eucharist (168*, 181*, 270*), nor about the sacraments of confirmation, matrimony, holy orders and extreme unction (how different from Poussin's approach in his almost contemporary paintings of *The Seven Sacraments*!);[48] for reasons apparent in Letter IX of the *Provincial Letters*, he has nothing to say about the Virgin Mary.[49] What is immediately obvious about Pascal's approach to religion is the stern emphasis he lays upon its moral dimension. A Probabiliorist himself, he castigated probabilism and the potential laxity of the Jesuits' ethical system. Underlying this sternly moralizing outlook is the undeviating belief that a man's attitude to the possibility of life after death will affect the moral quality of the life he lives upon earth:

> it is beyond question that the soul is either mortal or immortal; that must make a complete difference in matters of morality, and

> yet the philosophers have conducted their discussions of morality quite regardless of that (612*).[50]

Pascal lived his own life in this light, being pure and austere in his private conduct, an indomitable fighter for causes which he believed to be just, and a man who in all things tried fearlessly to face up to the truth. And yet it is impossible to resist the charge that he was somewhat arrogant, with the arrogance of a person who in so many respects (especially those pertaining to geometry) *knows* that he is right, though few others can perhaps understand him. Arrogant towards Jacques Forton in the dispute about the Holy Trinity; arrogant towards Noël in the dispute about vacuums; arrogant in the cycloid competition, especially towards La Loubère; arrogant towards his sister Jacqueline in the dispute over her bringing a dowry to Port-Royal; arrogant towards the priests of Port-Royal in the matter of signing the Formulary; arrogant, Nicole believed, in his bullying approach to the unbeliever in the plan of the *Thoughts*;[51] and, lastly, perhaps arrogant towards the Jesuits in that he unwarrantably travestied their ethical teaching (though this I doubt). There is little in his religious writings about reconciliation although much about justice; because of his preoccupation with the moral life he little heeds the teaching of Jesus 'that they may be one' (John XVII 22); from a reading of the *Provincial Letters* it is easy to conclude that there was, in 1656, a monster within the Church which had to be destroyed – just as, for Voltaire a century later, the monster to be destroyed had become the Church itself (in its then institutional forms). All too often, but mainly for the *moral* reason, Pascal busies himself with zealous sectarian infighting: hence Claudel's view that he has exerted a pernicious influence upon the course of Christian belief. 'Ever since the *Thoughts* were written', wrote Claudel to Gide in 1911,[52]

> people imagine that religion is a matter of sectarian fanaticism, that you have to shut your eyes, fall ill, cower in some corner, and amputate three or four of your human faculties, the noblest of them included. In fact, religion is as vast as the starry sky . . . On the contrary, it is the unbeliever who merely inhabits a cramped and diminished world.

The most startling aspect of Pascal's *moralistic* attitude is that, although he sets such supreme store by the dignity of man as a free moral agent, he shows so little awareness of secular, or religious, fellowship. Although the *Thoughts* imply that we should lead our lives on the moral basis that there may be a Hereafter, that book contains virtually no moral teaching. The situation is, of course, rather different in the *Provincial Letters,* especially with Pascal's emphasis upon contrition rather than attrition. Nicole, very early on in the history of attitudes towards the *Thoughts,* voiced disquiet about this lack of a sense of community.[53] Virtually the only *moralistic* reference in the latter work is the remark that 'there is a great difference between knowing about God and loving Him' (377*), a thought which recalls Pascal's root-and-branch hostility, in the *Provincial Letters,* to any mechanical observance of moral precepts. *Moralistic* is no doubt an inadequate word to convey the tone of Pascal's outlook in these matters. The tone is not exactly 'moralizing' though it occasionally comes fairly close to that in the *Provincial Letters*; 'moral' is also an unsatisfactory term because any ethical discussion of a problem could incline towards a position of immorality or amorality; and overall there is the further ambiguity of the French term *moraliste.* For upon Pascal's qualities as a *moraliste,* upon his understanding of human nature, depends so much of his appeal to the reader – especially when he is writing his vast work of Christian apologetics. Herein, as previously discussed, lies one of the most contentious aspects of the *Thoughts*: how valid is Pascal's analysis of human nature? 'All men naturally hate one another', he writes in *Thought* 210*: is this perception of human nature, so different for example from Rousseau's, actually true? It is not an entirely convincing rejoinder to argue that Pascal is speaking of man as yet unregenerated by baptismal grace.[54] For why should the world consist, today, of fallen men and women so full of hatred for one another? And are some of the baptized less full of hatred than some who have not been baptized? And why did God will this world of good and evil? Why the Elect and the damned? Why the system of human existence overall? Pascal does not really address these matters. He does not venture into the realms explored by Leibniz in his *Theodicy,* which also ranks, in its way, as one of the major works of religious apologetics. 'The great puzzles of existence', writes Peter Atkins,[55] ' – purpose and the existence of evil – are mere human inventions that have psychological and ethological origins'. It is fair to confront Pascal with this challenge from another scientist, though the world-views of the two men are three centuries apart. For both such views have to be considered by any unbeliever making up his mind about the validity of religion. Pascal is not merely the scientist who died in 1662 leaving behind him a quantity of religious jottings and an obscure handful of acknowledged publications. His is a system of thought inspiring many others', and on to which many graft their own. And central to his system of thought is the notion of purpose.

Hence the urgency of so much that he wrote. His combination of purposefulness, zeal sometimes disguised as ironical detachment, and even arrogance informs all his non-mathematical writing, generating its brilliance. It was Pascal (said Voltaire)[56] who, in the *Provincial Letters*, stabilized the French language, setting an enduring example of linguistic precision and accuracy.[57] Chateaubriand[58] and Renan[59] considered him a writer of matchless French prose. As a writer of satire he stands equal to Plato in his *Euthydemus*, or to Archilochus and Lucian. As a writer of impassioned invective he is on the level of Swift, Demosthenes and Juvenal; his delicious irony, seen to perfection in the discussions between the Jesuit priest and the narrator of the *Provincial Letters*, greatly influenced Edward Gibbon.[60] Pascal's influence upon the French language has also been immense: his writing is great precisely because he distrusted and shunned the adornments of preciosity. The formal elegance of his literary achievement in the *Provincial Letters* is an object of wonder: the more one knows of its complex and tangled background of casuistry and the theology of grace, the more one admires the adept and skilful manner in which he has simplified, without over-simplifying, those issues. The beauty of his treatment of the Jesuits in no way impairs its truth. No one could have been more penetrating than he was in his exposure of duplicity and folly.

> 'May the Jesuits kill the Jansenists?'; (738; *115*)
>
> 'Oh Father! . . . were all those men Christians?'
> 'What do you mean, Christians? . . . They are the only men through whom we govern Christendom today' (714; *86*).

The roll-call of Jesuit casuists, *de Graphæis, Squilanti, Bizozeri, Bisbe, Iribarne, Binsfeld* . . ., the changing tempo of the attack, the trivialization of the Virgin Mary, the ways in which the reader's attention is caught and held, the very invention of the Jesuit priest: all these are strokes of genius. Of all the bricks which make up Pascal's edifices few, if any at all, are 'new'. 'Pascal's *Thoughts* we shall never understand', writes Walter Pater,[61] 'unless we realize the under-texture in them of Montaigne's very phrases' – and not only phrases but concepts, as for example with *abêtissement*.[62] Nor are the quotations from the casuists 'new': they had recently been published, in more or less verbatim form, in various books and pamphlets. Even one of the best known sentences in the *Thoughts*, that the universe is a sphere 'whose centre is everywhere and whose circumference is nowhere' (199*), is the reverse of new:[63] it occurs in the writings of St Bonaventura[64], Vincent de Beauvais[65], Jean de Meung's portion of the *Roman de la Rose*,[66] Gerson[67] and Rabelais,[68] and is also said to have occurred in Empedocles, St Augustine, Hélinand de Froidmont and Trismegistic literature; it was probably relayed to him by his study of Montaigne.[69] Voltaire, on the other hand, unjustifiably understates Pascal's originality, extolling Corneille's.[70]

Pascal does not seek to deny any such intertextual criticism, preferring to deflect it to his own advantage:

> Let no one say that I have said nothing new; the arrangement of the material is new. When playing tennis, both players use the same ball, but one is better at handling it (696*).

His scientific originality – whether in the Arithmetical Triangle named after him,[71] conic sections,[72] atmospheric pressure,[73] hydrodynamics and hydrostatics,[74] or the method of indivisibles[75] – is not of the high but rare order of Gauss in mathematics or Einstein in physics: 'Pascal presented his research with greater method than did his contemporaries, but his original contribution has often been overestimated';[76] even Newton was to make a confession[77] rather similar to Pascal's, and so indeed could most scientists. Standing on the gigantic shoulders of Pythagoras, Euclid, Cavalieri, Mersenne and Roberval, Pascal applied the 'notion of pattern, common enough in geometry', to number itself, thus taking 'a highly significant step in the history of mathematics'[78] and paving the way for the achievements of Johann Bernoulli, Euler and Cayley. By *rearranging*, he *innovates*. In his work on the probability calculus, on the other hand, there is much greater originality, but perhaps less elegance in the arrangement of data.

The same fact of *rearrangement* rather than *innovation* is true of the Wager argument, which in the final analysis is not strictly mathematical, and which no previous writer, during the two millennia in which it had been known and used, had ever quite expressed in the same way or harnessed to the same practical purpose. Though Leibniz is incorrect in his suggestion[79] that Cicero foreshadowed[80] Pascal's Wager argument, it was Plato, writing at an even earlier date than Cicero, who had likened the best of human theories to the 'raft' upon which a man sails across life: 'not without risk, as I admit, if he cannot find some word of God which will more surely and safely carry him'.[81]

However, the appeal – and some would say the magisterial authority, others the tantalizing and ludic riddle – of the *Thoughts* seems enhanced by the very fact that it is an unfinished work, the parts of which have not even been 'arranged' by Pascal in any 'new' way. 'The parrot wipes its beak, even though it is clean' (107*), 'great Pan is dead' (343*), 'Port-Royal is well worth Voltigerod' (909*): what part, if any, would these sayings have played in the ultimate work? The images of the thinking reed (200*) and the condemned cell (163*, 434*), the vision of man suspended between two infinites (199*), the *Thought* concerning *divertissement* (136*) – all these, whether 'new' or not, are unforgettably vivid, as haunting as the image of the fettered prisoners in the cave in Book VII of Plato's *Republic*. Additionally, however, there is the enigma (which so strongly appealed to the Romantic rediscoverers of the *Thoughts*) that no one quite knows what, where or even occasionally why anything would have been included in the completed work. Like the meditations of Novalis, the *Thoughts* have all the mysteriousness of Gothic ruins or a mound of fragmentary stones.[82] In Pascal's nephew Étienne Périer's words,[83] they could be regarded as 'a confused, disorderly, inconsequential heap, serving no useful purpose'. The uncle himself believed that he would need ten years of sound health in order to complete his Apologia.[84]

Pascal's work on conic sections, probability theory and the cycloid curve, his study of atmospheric pressure and his proof that a vacuum can exist within a vacuum, his demolition of the Jesuits' probabilism and rebuttal of any mechanistic ethical system: these things have entered into the very fabric of the modern way of thinking. In no way, for example, is it possible to disprove his findings on probability theory. Nor, within the field of the history of mathematics, can it be disputed that his work on the properties of the cycloid paved the way for the infinitesimal, and hence the integral, calculus. For the archæologist (as it were) of the history of European thought all these things – scientific and non-scientific achievements alike, but more especially the former – lie in various layers of that *terra firma* which is the world's intellectual inheritance. Only the *Thoughts*, fluid and ever-changing as the sea, remain a perennially live issue. How many conversions have they produced? How many do they still produce? What is their value as an apologetic, theological or philosophical document? 'If I had had the misfortune to be an unbeliever', wrote Henri Bremond,[85] 'I should still have remained one despite many, indeed most, of Pascal's arguments'. Upon Renan Pascal's arguments produced an identical effect.[86] Leibniz doubted their ability to convert the hardened unbeliever.[79] They did not convert Voltaire. Boileau, on the other hand, observed that 'they had converted some people',[87] whilst Bayle argued[88] that it was the extraordinary piety and humility of Pascal's life which had had a more unsettling effect on atheists 'than if a dozen missionaries had been let loose upon them'. But Bayle's view must be discounted to the extent that the piety of Pascal's life was hardly different from that of many missionaries. It is the *Thoughts* which so often have had an unsettling effect, and intentionally so.

Together with Plato, Rousseau, Marx, Nietzsche, Dostoievski and the Book of Job, Pascal, in the *Thoughts*, is one of the askers of great questions. He is, in Isaiah Berlin's terms,[89] a 'hedgehog',[90] rather than a 'fox': knowing one big thing rather than (like the fox) many smaller things. Unlike most men and women, and even unlike many other thinkers, he was increasingly dominated by one concept. The one big thing he knew – beyond all peradventure, so far as he was concerned – is that man cannot reasonably avoid confronting the possibility of an existence beyond the grave. This is the problem with which he challenges all readers of the *Thoughts*. He does not address that even more fundamental question: why engage in any human activity at all?

Nevertheless, despite the various inadequacies of an apologetical system which is not total, Pascal leaves an abiding impression upon many hearts but perhaps still more minds.

Notes and References

CHAPTER 1 INTRODUCTION

1. For full details of Pascal's ancestry see Fric, R. (1923).
2. Cox, C.M. (1926), 691.
3. Mill began to learn Greek at three and Latin at seven (Mill, J.S. (1873), 5–9). Étienne Pascal's educational methods differed from James Mill's in that Étienne would not teach his son Latin until he was twelve years old (4).
4. And perhaps earlier still, in view of the interest which Mersenne shared with him in the properties of sound.
5. Mesnard, J. (1963).
6. Henry, C. (1884).
7. Rouse Ball, W.W. (1920), 308–10.
8. Pascal thus relied, like his predecessors, on Euclid's Parallel Postulate.
9. Only a few copies of the *Essay on Conic Sections* were published.
10. de Robillard de Beaurepaire, C.-M. (1902).
11. François, J. (1653), 22. Tallemant des Réaux, G. (1960–1), II 58 says '400 *livres*', in other words, about £2000 at today's prices.
12. The number of finished products was probably between twelve and thirty; the Pascalian calculator was generally manufactured according to the customer's personal specifications. Christiaan Huygens bought one, as did Leibniz, and two were bought by Marie-Louise de Gonzague, the wife of King John Casimir of Poland. Pascal gave one to Carcavi.
13. Concerning the Forton (or Saint-Ange) affair generally, see Jovy, E. (1927), Julien-Eymard d'Angers (1954), 147–8, 163 and Descrains, J. (1985), I 376–7.
14. The number of experiments carried out by or on behalf of Pascal has been a matter of speculation. Boyle, R. (1666), 4–7; Thurot, F.-C.-E. (1869), 19–20.
15. Gassendi, P. (1659), 214–16; Gassendi to Bernier, 6 August 1652. However, Gassendi credits Roberval with the original attempt at this experiment.
16. Knowles Middleton, W.E. (1964), 38–54.
17. E.g., Descartes, R. (1952), 620–1, *The Principles of Philosophy* [1644].
18. Étienne Noël to Blaise Pascal, October 1647 (1438–42); Blaise Pascal to Étienne Noël, 29 October 1647 (quoted in *New Experiments Concerning Vacuums*, 370–7); Étienne Noël to Blaise Pascal, November 1647 (1442–52).
19. Étienne Pascal to Étienne Noël, March–April 1648 (**II 255–79**).
20. Pascal to Fermat, 27 October 1654. He was present, for instance, at a meeting of this circle on 24 October 1654 (90).
21. Pascal's unfinished (and until 1663 unpublished) 'Fragment' of a proposed 'Treatise on Vacuums' underlined the principle of weather-forecasting based on observations of changes in atmospheric pressure (462–7).

22. This episode may have inspired *Thought* 579*, though it is not certain that Pascal ever did suffer such a mishap.
23. Kierkegaard specifically drew this parallel from Acts IX.
24. Claudel, P. (1965), 1009.
25. Kierkegaard, S.A. (1900–48), X (5), 19 (A 17) [1852].
26. de Condorcet, M.-J.-A.-N. C. (1803), I, xvi.
27. *Pensées*, ed. L. Lafuma, 1962, 373. One of these *Memorials* was on paper, the other on parchment; plate 4 depicts the paper version. The one on parchment was transcribed by Louis Périer, Pascal's nephew (Lafuma, L, (1962), 304).
28. *Circa* 10–21 January 1655, mid-January 1656, *circa* 30 November-4 December 1656, May 1658, August 1659. The first two visits are the only definite ones, but the visit in 1658 would have been the time when he expounded *Thought* 149*. Mesnard, J. (1951).
29. Pascal was probably the author of this work although the attribution has not been definitively established. The manuscript was found amongst Jacqueline Pascal's papers after her death in 1661.
30. However, the attribution of this work to Pascal is not totally certain.
31. Or perhaps as late as the spring of 1658.
32. Lancelot, C. and Arnauld, A. (1660), 23–5, 'D'une Nouvelle manière pour apprendre à lire facilement en toutes sortes de langues'. See 1454–5 for an extract from Chapter VI of the *Port-Royal Grammar*.
33. Jacqueline Pascal to Pascal, 26 October 1655. For a discussion of the content of the *Port-Royal Grammar* see Miel, J. (1969b).
34. Arnauld, A. (1667), 1–18.
35. Arnauld, A. and Nicole, P. (1664), 108–24.
36. Church, A. (1958); Miel, J. (1969b); James, E. D. (1972).
37. E.g., *Port-Royal* (1818). This was a translation of Arnauld, A. and Nicole, P. (1664).
38. Pascal to François Le Pailleur, *circa* February 1648.
39. Ernst, P. (1989), 132–48, 188–90.
40. However, Goldmann, L. (1955), Marin, L. (1975a, 1975b) and Morot-Sir, É. (1977) all dispute that Pascal planned to write an apologia. To Chevalier, J.-L.-A. (1934), Hubert, M.-L. (1952), Mantoy, J. (1955), Ernst, P. (1970), Ernst, P. (1989), Laberthonnière, L. (1903), Roberts, D. E. (1957), Lacombe, R.-E. (1958), Magnard, P. (1980), Mesnard, J. (1976) and Pugh, A. R. (1984), on the other hand, the defence of the Christian religion is central to the work. Admittedly, Pascal does not specifically use the word 'apologia' in the *Thoughts*.
41. [Georges Pirot's] *Defence of the Casuists* had appeared in December 1657.
42. As reported by Marguerite Périer.
43. Though essentially one *Letter*.
44. Roberval established that the area of a cycloid was $3\pi r^2$ (194). This was probably as early as 1634. But much of the theoretical basis of his work remained unknown during his lifetime.
45. According to Ward, J. (1740), 97, Pascal used the alias Jean de Montfert when issuing this challenge to Wren. A copy of Wren's attempt at a solution of Jean de Montfert's problems is preserved in the heirloom copy of the Wren family's *Parentalia* (Whiteside, D. T. (1960), 111).

46. On the controversy which ensued between Pascal and La Loubère, see Tannery, P. (1926).
47. Ward, J. (1740), 97. See also Wren, C. (1750), 242–3.
48. Pascal's other definite alias, Salomon de Tultie, is a further anagram of Louis de Montalte. It is used in reference to himself in *Thought* 745* and may have been intended as the name on the title page of his apologia for the Christian religion. The possible alias Jean de Montfert would evidently have been an allusion to Cler*mont*-*Fer*rand.
49. On the evidence of Newton, I. (1959–77), I *passim* it would seem that Newton was not. See likewise Newton, I. (1967–81), I, 13 and VII, 169, n. 21.
50. Aiton, E.J. (1985), 57; Hofmann, J. E. (1974), 179–80; Child, J. M. (1920), 196, n. 4: 'There was absolutely nothing in Pascal to suggest the sign or the rules for *differentiation,* and Leibniz might just as easily have obtained his ideas on integration from Galileo or others as from Pascal.'
51. Greenstreet, W. J. (1927), 158–9; Westfall, R. S. (1993), 39, 98.
52. Cavalieri, B. (1635). See Russo, F. (1962).
53. See Taton, R. (1962b).
54. Oldenburg, H. (1965), 225; Henry Oldenburg to M. Saporta, 6 May 1659. Oldenburg had arrived in the French capital by 22 March 1659 (Oldenburg, H. (1965), 208).
55. Pascal to Fermat, 10 August 1660.
56. See **X 318–19** and Mesnard, J. (1952b). A few of Pascal's theological books may still be in existence.
57. Huygens, C. (1888–1950), II 456–7; Pierre de Carcavi to Christiaan Huygens, 14 August 1659.
58. See Duclou, M. (1950).
59. *Pensées*, ed. L. Lafuma, 1962, 434.
60. Except for one possibly worn by Elizabeth I, there is no previous instance of the use of a wristwatch. The earliest watches were generally worn on a chain around the neck. In Pascal's day many were carried in pockets, which by then had become a standard feature of clothing. The next known instance of a wristwatch occurs in Switzerland in 1790.
61. **IV 77 n. 1**: 'There is a little boy about 12 years old who cannot read: I want to find out whether he can learn to do so by M. Pascal's method' (Antoine Arnauld to his niece Angélique, 31 January 1656).
62. Faugère, A.-P. (1845), 438.
63. See Gazier, A. (1911).
64. Pascal's autopsy revealed disorders of the stomach and adjacent organs, together with a serious lesion of the brain.
65. Racine, J. (1950–52), II, 78. Maire, A. (1925–7), V, 38–42. There is, however, a suggestion that the affair may have been more carnal in nature (Fléchier, V.-E. (1844), 87). On the possibility of Pascal's love affair with Charlotte de Roannez, see Faugère, A.-P. (1844), I, lxv–lxix. See also Chamaillard, E. (1923). The *Discourse on the Passions of Love* seems to have been written about 1653 or 1654; it is perhaps a symposium, or garland, of thoughts gathered together on the subject of love from a worldy circle in which Pascal moved. Even if it is by Pascal himself (which seems most unlikely), it would not point to any intimate

experience of love on his part, although it might have some reference to his alleged courtship in 1652.

66. Ojardias, A. (1910), 195–6.
67. For further details of its provenance, see Maire, A. (1925–7), V, 84.
68. The cast at Newnham College, Cambridge was presented to that institution by Mrs H. F. Stewart in 1948. Concerning death masks generally, see Benkard, E. (1929), in which reference is also made to Pascal's (page 69).
69. One of these formerly belonged to Julien Green (Green, J. (1951), 338; diary entry for 30 January 1950).
70. Ojardias, A. (1910), 195–204; Dorival, B. (1962); Gazier, A. (1909), 14.
71. Dorival, B. (1962), 93–4, 96 suggests rather improbably, however, that it was drawn at some time between 1677 and 1681. The Domat drawing formerly belonged to Maurice Barrès and is now at the Bibliothèque Nationale.
72. Or perhaps 1663: see Dorival, B. (1962), 94–5. In 1914 (**X 318**) and in 1927 (Maire, A. (1925–7), V, 85) this picture belonged to a descendant of a cousin of Pascal's on his mother's side. In this portrait Pascal looks to the right.
73. This earlier engraving by Gérard Édelinck can be seen in Perrault, C. (1697–1700), I, facing 65. Dorival, B. (1962), 97.
74. This is illustrated on the front cover of *Pensées*, ed. L. Lafuma, 1962. In this portrait Pascal looks to the left.
75. See Gazier, A. (1909), 14.
76. This is one of the two bought by the Queen of Poland. It was transferred from Warsaw to Dresden, probably during the second phase of the dual monarchy from 1709 to 1763.
77. The French inch in 1645 was equivalent to the current 0.419 cm, or to 1.065 of the twentieth-century English inch.
78. See Payen, J. (1963). The specimen now belonging to IBM was acquired in 1942 from an antique-dealer who had bought it from a vendor who believed it was a musical box!
79. Racine, J. (1950–2), II, 153.
80. Nicole, P. (1730–5), VIII, 243; Pierre Nicole to Marquis Renaud de Sévigné [*circa* 1672]: James, E. D. (1972), 111.
81. In 1647 or 1648 Pascal suffered temporary paralysis of the legs (37; **IX 321**).
82. **IX 202 n. 1**; Ismaël Boulliau to Leopoldo de' Medici, 13 June 1659 (Huygens, C. (1890), III, 466).
83. Pascal to Fermat, 10 August 1660.
84. According to a story which first surfaced 75 years after Pascal's death (Boileau, J.-J. (1737), 207). See Onfray, R. (1949), 5–9.
85. Admirably analysed in Jerphagnon, L. (1962). See also Jordan, H. R. (1909).
86. One of the world's 300 geniuses, according to Cox, C. M. (1926), 690–3. Pascal is reckoned to have had an intelligence quotient of 180.
87. Périer, É. (1670), *Pensées*, ed. L. Lafuma, 1962, 16.
88. Arnauld, A. and Nicole, P. (1664), 341.
89. It is a fact that the letters become longer as the controversy continues, Letter XVI being the longest of all.

90. Filleau de La Chaise, J. (1672). See McKenna, A. (1988).
91. See A List of Pascal's Writings (pp. 277–82).
92. **IX 202 n. 1**; Ismaël Boulliau to Leopoldo de' Medici, 13 June 1659 (Huygens, C. (1890), III 466).
93. 'It can be said in all truth that we have lost perhaps one of the greatest minds that has ever existed. I see no one who can compare with him. . . He will be little heard of by posterity, for those of his works which we still possess are incapable of giving a true impression of the vast range of his mind. . . Whereas all that remains of that great mind are two or three little works including some very useless things. . . ' (Maire, A. (1925–7), V, 297; Pierre Nicole to M. de Saint-Calais, 3 September 1662). For reaction to Pascal during the period prior to the publication of Voltaire's *Philosophical Letters*, see McKenna, A. (1990).
94. Perrault, C. (1697–1700), I 65–6. Pascal's appearance in Perrault, C. (1697–1700), seemingly by way of an afterthought, is explained in Dorival, B. (1962), 96–7. His entry was tipped in immediately after that of the eminent Byzantinist Charles du Cange, which has identical pagination; he is not mentioned in the index to volume I. Why he was not included in the second volume is unclear.
95. Quoted in Taylor, F.A. (1943), xxvii; Voltaire to Cideville and Formont, June 1733. For Voltaire's evolving, and gradually more favourable, attitude to Pascal, see Waterman, M. (1942).
96. See Finch, D. (1940); Vamos, M. (1972); Barker, J. (1975).
97. de Condorcet, M.-J.-A.-N. C. (1847–9), IV, 292, *Notices Inserted by Condorcet in the Complete Edition of the Works of Voltaire* [1785].
98. de Chateaubriand, F.-R. (1802); de Maistre, J.-M. (1821), I 522; Sainte-Beuve, C.-A. (1848–59); Cousin, V. (1857).
99. For the general reaction to Pascal's *Thoughts* since the year of their first publication see Amoudru, B. (1936). More specifically, for the French intellectual reaction to them in the nineteenth century see Renouvier C.-B. (1896–7) and for the reaction in France between 1842 and 1942 see Francis, R. (1959).
100. de Condorcet, M.-J.-A.-N. C (1803), II, 285.
101. For the nineteenth- and early twentieth-century intellectual reaction to Pascal's thought generally, see Eastwood, D.M. (1936).
102. Gassendi to Barancy, 13 December 1648.
103. Nicole, P. (1730–5), VIII 242; Pierre Nicole to Marquis Renaud de Sévigné [*circa* 1672].
104. Bayle, P. (1715), III 138 [1696].
105. von Leibniz, G.W. (1849–60), III 72–3 n.; Leibniz to Jakob Bernoulli, April 1703 (*deleted postscript*).
106. Voltaire (1734), Letter XXV.
107. de Vauvenargues, L. de C. (1746), 376.
108. Hume, D. (1882), II, 304.
109. de Chénier, A.-M. (1914), 183.
110. de Chateaubriand, F.-R. (1802), III, 66.
111. von Schlegel, K.W.F. (1815), II, 181 [1812].
112. Macaulay, T.B. (1832), 274.
113. Disraeli, B. (1844), I, 253. Julien Green has spoken of Pascal in identical terms (Green, J. (1951), 338; diary entry for 30 January 1950).

114. Sainte-Beuve, C.-A. (1848), III, 261. See Stewart, H.F. (1915), 94.
115. Comte, A. (1894), 12. In this calendar, issued by Comte in 1849, Pascal's day is 16 October, in the week of Bacon in the month of Descartes.
116. de Goncourt, E. and J. (1956), I, 1272, I, 1367 (11 May and 21 December 1863).
117. de Goncourt, E. and J. (1956), I, 1367 (21 December 1863).
118. Eliot, G. (1871), I, 41.
119. Church, R.W. (1895), 24. From one of a series of sermons on 'Companions for the Devout Life' preached at St James's Church, Piccadilly, London in 1875.
120. Jowett, B. (1899), 107. From a sermon preached in Balliol College Chapel on 22 May 1881.
121. von Nietzsche, F.W. (1930), III, 159 [1881].
122. von Nietzsche, F.W. (1930), VI, 39 [1887].
123. von Nietzsche, F.W. (1932), VIII, 497; Nietzsche to Georg Brandes, 20 November 1888.
124. Daudet, A. (1931), 51 [1889].
125. Pater, W.H. (1895), 178.
126. James, W. (1902), 286.
127. Stephen, L. (1898), II 243.
128. Claudel, P. (1949), 184; Claudel to Gide, 7 December 1911.
129. Strachey, G.L. (1912), 58.
130. Barrès, A.-M. (1918), 117.
131. Bourget, P.-C.-J. (1922), I, 213.
132. Huxley, A.L. (1929), 212, 236.
133. Carré, J.-R. (1935), 10.
134. Maurras, C.-M.-P. (1953), 20.
135. Duhamel, G. (1941), 175, 176.
136. Russell, B.A.W. (1945), 768.
137. Collingwood, R.G. (1946), 80.
138. Green, J. (1951), 261; diary entry for 25 May 1949.
139. Knox, R.A. (1950), 224.
140. Fletcher, F.T.H. (1954), 74.
141. Roberts, D.E. (1957), 15.
142. Hollis, M.C. (1968), 97.
143. Cupitt, D. (1980), 7.
144. Hay, M.V. (1962), 13.
145. Ephesians I, 23.

CHAPTER 2 FOUNDATIONS

1. Few copies of the *Essay on Conic Sections* were published; two are known to have survived. Leibniz, finding two amongst Pascal's papers in 1676, kept one for himself (65).
2. Descartes, R. (1897–1909), VI, *Geometry, passim* and especially Book III (442–85) [1637]. Using algebraic methods, Descartes elaborated a

completely general theorem of conics of which Apollonius's propositions were merely examples.

3. Desargues had preceded Pascal in his study of conics. The seminal nature of Desargues's contribution to mathematics did not become fully clear until the rediscovery of some of his writings in 1846 and again in 1950. Taton, R. (1951).
4. Preceded by Pratt's (1616) and Schickard's (1623) devices, it was adapted by Morland (1663), Cotterell (1667), Leibniz (stepped reckoner: 1673), Grillet (1678), Browne (1700), Claude Perrault (1700), Poleni (1709), Lépine (who considerably simplified it: 1725), Braun (1727), Leupold (1727), Hillerin de Boistissendeau (1730), Pereire (1750), Hahn (1774), Stanhope (1776) and Müller (1783). Pascal's type of calculating-machine reached its zenith in Thomas de Colmar's arithmometer (1820), which was the first reliable mass-produced mechanical calculator. Turner, A.J. (1987), 167, 286.
5. For its method of operation see in particular Fric, R. (1923), 37–41.
6. This was described in some detail in Diderot, D. and d'Alembert, J. Le R. (1751–65), I 680–4.
7. This was not a clock but was given that name because it was a device with moving mechanical parts, including wheels; it was probably based on Napier's Bones.
8. Goldstine, H. H. (1972), 7.
9. Birch, T. (1757), 73.
10. The first *operational* difference engine was manufactured in 1843 by Georg and Edvard Scheutz.
11. But this is not to underestimate the influence of Pascal's Triangle upon the development of the computer.
12. Gassendi and the other atomists believed in an interatomic void by means of which they explained rarefaction, condensation and motion; but they denied that it was possible to produce a perceptible void.
13. Descartes, R . (1952), 612–15, 620–1, *The Principles of Philosophy* [1644].
14. But he came round to agreeing with Pascal that the phenomena attributed by Galileo to Nature's abhorrence of vacuums were actually due to atmospheric pressure.
15. Aiton, E. J. (1972); Scott, J.F. (1952), 167–81.
16. Jacqueline Pascal to Gilberte Périer, 25 September 1647.
17. Étienne Noël to Blaise Pascal, October 1647 (1438–42); Blaise Pascal to Étienne Noël, 29 October 1647 (quoted in *New Experiments Concerning Vacuums*, 370–7); Étienne Noël to Blaise Pascal, November 1647 (1442–52).
18. Étienne Pascal to Étienne Noël, March-April 1648 (**II 255–79**).
19. Descartes, R. (1952), 1289; Descartes to Marin Mersenne, 13 December 1647.
20. Descartes, R. (1897–1909), V, 366; Descartes to Pierre de Carcavi, 11 June 1649.
21. Descartes, R. (1897–1909), V, 391; Descartes, R. (1952), 1289, n. 1; Descartes to Pierre de Carcavi, 17 August 1649.

22. See Jovy, E. (1928).
23. Rohault, J. (1676), I, 100.
24. Rochot, B. (1963), 55.
25. This work, published in October 1648, a month or so after the experiments on the Puy de Dôme, also contains the letter from Florin Périer (395–9) informing Pascal of their success.
26. Pascal understates the radius of the earth by approximately 20%.
27. Μάθησεως (*máthêseôs*) is the genitive form of the Greek noun μάθησις (*máthêsis*) and means 'of mathematics or science'. As such it occurs in the 4th *Rule for the Direction of the Mind* (Descartes, R. (1897–1909), X 375–7 (Descartes, R. (1952), 49–50), where it basically means 'order' or 'method', the foundation of that 'method' being mathematical. *Máthêsis* also formed part of the title of one of Wallis's mathematical works in 1657.
28. These, it seems, were the informal scientific and learned circles presided over by Henri-Louis Habert de Montmor, Jacques Dupuy, Jacques Rohault, Pierre Petit and perhaps others: Brown, H. (1934). The Montmor academy had originally been Mersenne's (1635–48) and was later chaired by François Le Pailleur, with help from Claude Mylon, and finally by Montmor: Mesnard, J. (1963). The Montmor academy gradually evolved into the Académie Royale des Sciences, which was set up in 1666, four years after the Royal Society (of London) received its charter; these were soon to be followed by the scientific societies of Florence and Berlin. Pascal's way of addressing these informal academies may have been a fairly standard one: Oxford University was described as 'celeberrima Academia Oxoniensi' on the title page of Wallis, J. (1653).
29. Edwards, A. W. F. (1987), 86, n.16. In 1675–6 Leibniz, with his assistant E.W. von Tschirnhaus, copied and made notes from Pascal's unpublished mathematical papers (Hofmann, J.E. (1974), 178–81; Aiton, E.J. (1985), 56).
30. But mentioned in the *Treatise on the Equilibrium of Liquids* (416). This was probably a generalized formulation of Torricelli's two laws of stasis (concerning the lever and inclined planes) outlined in his *Geometrical Works*.
31. Pascal's 'Conical Contacts' and 'Solid Loci' were last studied, by Leibniz, towards the end of his four-year stay in Paris (63–5, **II 220–4**: Leibniz to Étienne Périer, 30 August 1676). The material on conic sections was handed over to Leibniz in January 1676. Hofmann, J. E. (1974), 179–80; Aiton, E.J. (1985), 56. Together with all the other mathematical manuscripts of Pascal borrowed by Leibniz from Pascal's nephews Étienne, Louis and Blaise Périer (and returned to them on 30 August 1676), 'Conical Contacts' and 'Solid Loci' had vanished by the time of the Abbé Gilles Filleau des Billettes's letter to Leibniz dated 28 May 1697 (**II 220**).
32. Descartes, R. (1897–1909), VI, *Geometry*, 377–87 [1637].
33. I.e., by generalization.
34. Pascal may partially have completed 'The French Apollonius Improved Upon' and 'Spherical Contacts'. All that we know of

these is contained (82) in Pascal's letter to Fermat dated 29 July 1654 (77–83); there is nothing in the Leibniz papers. 'The French Apollonius Improved Upon' seems (73) to have been inspired by Viète, F. (1600), the latter's critique of the mathematical work of Apollonius of Perga.

35. Concerning plane loci, the science of perspective and gnomonics. The science of sundials was an interest which Pascal shared with Desargues and with the latter's other major disciple, Philippe de La Hire. Pascal also refers tantalizingly to 'innumerable and varied research topics which I have fairly well in hand' (74).
36. Pascal's investigations of the properties of magic squares were almost certainly utilized in Arnauld, A. (1667), 325–45 (1388–1400).
37. 'Generatio Conisectionum, Tangentium et Secantium; seu Projectio Peripheriæ, Tangentium et Secantium Circuli, in quibuscumque Oculi, Plani ac Tabellæ Positionibus' ('The Generation of Conic Sections, Tangents and Secants; or the Projection of the Circumference, Tangents and Secants of the Circle, for Every Position of the Eye and of the Plane of the Figure').
38. Hofmann, J. E. (1974), 180; Aiton, E. J. (1985), 56. See also **II 217–33**.
39. de La Hire, P. (1673), de La Hire, P. (1679).
40. It would seem that the *Treatise on the Arithmetical Triangle*, with its two appendices, was written at about the same time as *The Probability Calculus*: between July and September 1654. Edwards, A.W.F. (1987), 58 suggests August of that year.
41. The most conveniently available version of the Pascal–Fermat correspondence is in **III 369–427**.
42. 29 July 1654 (77–83), 24 August 1654 (84–9).
43. See on this subject David, F.N. (1962), 81–97, 'The Arithmetical Triangle and Correspondence between Fermat and Pascal'.
44. David, F.N. (1962), 89 n. ‡ comments: 'The Chevalier de Méré was obviously such an assiduous gambler that he could distinguish empirically between a probability of 0.4914 and 0.5., i.e. a difference of 0.0086, comparable with that (0.0108) of the gambler who asked advice of Galileo.'
45. Concerning the dice problem, see Ore, O. (1960).
46. Edwards, A. W. F. (1982) and Edwards, A. W. F. (1987), 58–60, 71–9, 138–49.
47. Keynes, J. M. (1921), 82.
48. 'It is doubtful on the basis of the mathematics present in the letters exchanged that either man would have been able to compute without tables the number of outcomes unfavourable to a player because of the order of wins' (Mahoney, M. S. (1973), 398).
49. Letter **iv** not being a reply to letter **iii**.
50. David, F. N. (1962), 96 concludes from this, and from the break in their correspondence, that Pascal may have realized that 'he was of inferior intelligence, mathematically, to Fermat'. It is certainly true that Pascal brought the discussion to an abrupt end on 27 October 1654 (90).
51. Huygens, C. (1888–1950), XIV, 3–179, and especially 60–90, 'On Calculation in Games of Chance' [1656–7].

52. This consisted of the actual *Treatise on the Arithmetical Triangle* (97–107) plus separate though related publications:- four papers (108–29) on various applications of the Triangle (including the probability calculus); and four sets of papers on the powers of integers (130–9), orders (140–7), combinations (148–58) and multiple numbers (159–71) respectively. Edwards, A. W. F. (1987), 58–68, 70–84.
53. Aiton, E. J. (1985), 18–21, 44–5. In his use of the additive process to determine binomial coefficients Pascal anticipated the work which Leibniz was to do in 1666 in his *Dissertatio de Arte Combinatoria.*
54. On Pascal's Triangle see Struik, D.J. (1969), 21–6.
55. Edwards, A. W. F. (1987), 87.
56. de Condorcet, M.-J.-A.-N. C. (1795), 289–90 [1794].
57. Walker, H. M. (1929).
58. As early as 1725 Abraham De Moivre published a manual on life-annuities.
59. Fry, T. C. (1928).
60. The importance of the notion of probability in Arnauld, A. and Nicole, P. (1664) has been overstated in Hacking, I. (1975), 73–84.
61. For an outline of *The Summing of the Powers of Integers* see Marie, M. (1883–8), IV, 188–9, Boyer, C. B. (1939), 148–50 and Edwards, A. W. F. (1987), 82–4. In this tract Pascal employed the instance $\sum_{1 \leq i \leq 4} (3i+2)^3$, but without recourse to algebra.
62. Edwards, A. W. F. (1987), 82.
63. Pascal to Pierre de Fermat, 29 July 1654.
64. Concerning the notion of *honnêteté* see Magendie, M. (1925).
65. de Méré, A. G. (1700), 4–5, 8, 19, 29–30. Méré died in 1685.
66. Koyré, A. (1972), *passim.*
67. Romans I, 20: *The invisible things of him from the creation of the world are clearly seen, being understood by the things that are made, even his eternal power and Godhead.*
68. Ephesians I, 23.
69. Vaughan, R. A. (1856), I, 191.
70. van Ruysbroeck, J. (1912), 19.
71. 2 Timothy I, 12.

CHAPTER 3 JANSENISM

1. Knox, R. A. (1950), 210.
2. Descartes, R. (1952), 166, *Discourse on Method* [1637].
3. Descartes, R. (1952), 157–62, *Discourse on Method* [1637].
4. Clark, R. (1932), 61.
5. Later to become Arnauld, A. (1667).
6. Clark, R. (1932), 60–2.
7. Knox, R. A. (1950), 178.
8. Mortimer, E. (1959), 54.
9. Mortimer, E. (1959), 58.
10. Knox, R. A. (1950), 196.
11. Knox, R. A. (1950), 198.

12. Knox, R.A. (1950), 189.
13. Knox, R.A. (1950), 196.
14. In 1608 and in 1636.
15. *Ex Omnibus Afflictionibus*, 1 October 1567.
16. *Provisionis Nostræ*, 29 January 1579.
17. *In Eminenti*, 6 March 1642.
18. Denzinger, H.J.D. (1957), 1092–6.
19. Parish, R.J. (1989), xii. Only the first Proposition, which is admittedly the crucial one, appears – almost *totidem verbis* – in the *Augustinus*, 3 xiii, col. 334.
20. *Cum Occasione Impressionis Libri*, 31 May 1653. The precise texts condemned by Papal decree are to be found in Denzinger, H.J.D. (1957), 1092–6.
21. Brief of Pope Innocent X, 20 September 1654.
22. Knox, R.A. (1950), 197.

CHAPTER 4 A 'GOD-INEBRIATED MAN'

1. The phrase was used of Spinoza in Novalis (1953–7), II, 476, although the God of Spinoza is quite different from Pascal's.
2. The *Conversation with M. de Saci regarding Epictetus and Montaigne.*
3. On the *Conversation with M. de Saci* generally, see Courcelle, P. (1960).
4. On Pascal's attitude towards Montaigne, see Barnett, R.L. (1986), Bart, B.F. (1955–6), Brunschvicg, L. (1944), Chambers, F.M. (1950), Croquette, B. (1974), Friedenthal, R. (1969), Grote, J. (1877), La Charité, R.C. (1973) and Uhlír, A. (1907).
5. Knox, R.A. (1950), 203.
6. Matthew XXVI, 38.
7. 'As an unclean beast for the mire'. Cf. Horace: *Epistles*, I, ii, 26: 'vixisset canis immundus vel amica luto sus' ('would have lived the life of an unclean dog or of a sow attached to its mire').
8. Leviticus, Isaiah, The Song of Solomon.
9. Ephesians, Philippians, I Corinthians, Hebrews.
10. Bremond, H. (1916–36), IV, 390.
11. Knox, R.A. (1950), 222.
12. Valéry, P. (1923), 170.
13. 2 Timothy I, 12.
14. John XIV, 6–7.
15. Perrault, C. (1909), 29.
16. Maire, A. (1925–7), II, 56.
17. The *Provincial Letters* remained on the Papal Index until the abolition of that (by now notorious) instrument of censorship in 1966. As late as 1848 it was an offence worthy of excommunication even to be the owner of a copy of the *Provincial Letters* within the archdiocese of Paris (Renan, J.-E. (1947–61), I, 305).
18. Maire, A. (1925–7), II, 57.
19. For a timetable of the events pertaining to the writing and publication of the *Provincial Letters* see Parish, R.J. (1989), xi-xviii.

20. Baudry de Saint-Gilles d'Asson, A., (1936), 171; 3 April 1656.
21. Denzinger, H. J. D. (1957), 1098.
22. Denzinger, H. J. D. (1957), 1099. Towards the end of his pontificate Alexander published this Formulary in his constitution *Regiminis Apostolicis* on 15 February 1665.
23. It was first published by Bossut in 1779.
24. Sainte-Beuve, C.-A. (1840–61), III 110.
25. Malebranche, N. (1965), XII 177 and *passim* [1688].
26. Concerning the significance of miracles to Pascal see Shiokawa, T. (1977). Ernst, P. (1989) and Hammond, N.G. (1992) dispute the potential importance of miracles in Pascal's apologetical system.
27. See Gazier, A. (1909), pl. 129. The clouds almost resemble an eye, so much so that in Sainte-Beuve's view: 'he changed his seal, using as his *arms* not a *Sky* (people have been mistaken about this) but, something rather less attractive, an *Eye* surrounded by a Crown of Thorns' (Sainte-Beuve, C.-A. (1840–61), III, 115).
28. In 1651 Roannez, created in 1566, was the third oldest – after Thouars and Uzès – of the thirty-six non-Royal French dukedoms; it had originally been created in 1519 (de Barthélemy, É.-M. (1867), *passim*).
29. These are unfortunately lost.
30. Ezekiel XVIII 31.
31. Isaiah XLV 15.
32. Isaiah XLV 15: 'Vere tu es Deus absconditus'. Cf. *Thoughts* 781*, 427*, 446*, 149*.
33. Romans I, 20.

CHAPTER 5 THE *PROVINCIAL LETTERS*

1. Gide, A. (1951), 991; 23 June 1930.
2. On Pascal's attitude towards casuistry see Eliot, T.S. (1931). Perhaps because he felt such a close affinity with Pascal, this essay is one of Eliot's finest achievements.
3. Popes Paul V, Gregory XV and Urban VIII.
4. Pope Urban VIII.
5. Étienne Bauny (1564–1649) was a Jesuit professor of moral theology whose *Compendium of Sins* (1634) was condemned at Rome in September 1640 and in Paris in April 1642.
6. François Hallier (1595–1659), Bishop of Toul (1643–56) and of Cavaillon (1656–9), secured the condemnation of Bauny's book by the French bishops. He later sympathized with the Jesuits.
7. This is corroborated in Sánchez, T. (1615), I, 30–38 [1613]. The point is developed in Arnauld, A. (1775–83), XXIX, 74 [1643].
8. Bevan, E. R. (1932), 192.
9. John I, 29.
10. Instructions condemned on 4 March 1679 by Pope Innocent XI (Denzinger, H. J. D. (1957), 1201).
11. Bevan, E. R. (1932), 193.
12. Ephesians VI 12.

13. I.e., Annat, F. (1656).
14. John XVIII 36.
15. The church, or *temple*, of Charenton-le-Pont, five miles south-east of Paris, which Henri IV had allowed the Protestants to build and which was destroyed in 1685 at the time of the Revocation of the Edict of Nantes.
16. Whatever Pascal's deviations from the path of strictest accuracy may be, the general line of his argument, as illustrated by quotation, can scarcely be faulted.
17. Juan Azor, Niccolò Baldelli, Fernando de Castro-Palao, Gilles de Coninck, Estevão Fagundez, Vincenzo Filliucci, James Gordon, Jaime Granados, Gaspar Hurtado, Pedro Hurtado de Mendoza, Paul Laymann, Leonard Lessius, Juan de Lugo, Luis Molina, Fernando Rebello, Valère [Reginaldus], Manoel de Sá, Juan de Salas, Tomas Sánchez, Francisco Suárez, Francisco de [Toletus], Luis [Turrianus], Gregorio de Valencia and Gabriel Vasquez: anthologized in de Escobar y Mendoza, A. (1659) [1644]. They came from Spain, Portugal, Italy, Flanders, Germany, France and Scotland. Two were cardinals. All were Jesuits, and all were 'serious doctors' according to Pascal's (ironic) definition of the word. Hurter, H. (1871–86), I, and Maynard, M.-U. (1851), II 443–74.
18. de Escobar y Mendoza, A. (1659), 734 [1644].
19. This should read 'Praxis. . . ex Societatis Jesu Doctoribus', not ' . . . Schola' (Cognet, L. (1965), 98, n. 2).
20. 'If he leaves off his habit in order to steal, or to fornicate, in secret'.
21. 'So that he may go unrecognized to a brothel'.
22. Diana, A. (1629–41), III ii 115. Cognet, L. (1965), 98, n. 3. Cf. Diana, A. (1698), VII, 93 [1629–41].
23. Eighteen of these names are in varying degrees misspelt, including, for example, Achokier (i.e., de Cochier), Tambourin (i.e., Tamburini) and de Graphæis (i.e., Graffio).
24. Denzinger, H. J. D. (1957), 1190; decree of Pope Innocent XI, 4 March 1679.
25. Cognet, L. (1965), 139 n. 5. Cf. de Escobar y Mendoza, A. (1659), 367 [1644].
26. Cognet, L. (1965), 140 n. 1.
27. Lessius, L. (1606), 268–9 [1605]. Cf. Cognet, L. (1965), 141, n. 1.
28. Cognet, L. (1965), 110 n. 3. This instruction was condemned on 4 March 1679 by Pope Innocent XI (Denzinger, H.J.D. (1957), 1187).
29. Claude de Montrouge was a judge of the Supreme Court.
30. Criminal justice was dispensed from the Châtelet fortress on the right bank of the Seine, which was demolished in 1802.
31. Francesco Amico (1578–1651), an Italian Jesuit, Chancellor of the University of Graz.
32. This instruction was condemned on 24 September 1665 by Pope Alexander VII (Denzinger, H. J. D. (1957), 1117).
33. Accurately quoted from Amico, F. (1642), V xxxvi (118).
34. Caramuel de Lobkowitz, J. (1652–3), 543.
35. 'Sometimes he even has to kill'.

36. Cognet, L. (1965), 128, n. 4. Escobar y Mendoza, A. (1659), 119 [1644], citing Lessius, L. (1606), 88 [1605]; the reference to Molina, L. (1588) has not been established. The instruction that 'I can properly kill a thief to save a single gold piece' was condemned on 4 March 1679 by Pope Innocent XI (Denzinger, H.J.D. (1957), 1181).
37. Hurtado de Mendoza, P. (1631), II, xv (iii) 4 (48).
38. E.g., Matthew XIX 24, Matthew VI, 19–21 and the parables of the widow's mite (Mark XII 41–4) and of Dives and Lazarus (Luke XVI, 19–31).
39. E.g., Vasquez, G. (1618), 18–21.
40. Le Moyne, P. (1645), 621–5 [1640–3].
41. Le Moyne, P. (1645), 621 [1640–3].
42. Le Moyne, P. (1652), 149, 129, 127, 157, 163.
43. *Etre toujours au bal* is the phrase used in the original.
44. de Barry, P. (1655), 33, 143, 172, 420, 261–2, 59–60, 156.
45. de Escobar y Mendoza, A. (1659), 183 [1644].
46. 'Two halves constitute one mass'.
47. Bauny, É. (1646), 312 [1634]. This instruction was condemned on 4 March 1679 by Pope Innocent XI (Denzinger, H.J.D. (1957), 1203).
48. Azor, J. (1610–16), I, 631.
49. Cognet, L. (1965), 169, n. 3.
50. This allegation from Sirmond, A. (1641) was the deepest wound inflicted by Pascal upon the Jesuits. It was, for example, the subject of the fierce argument between Boileau and the Jesuit companion of Bourdaloue which took place on 5 January 1690 (de Sévigné, M. (1963–78), III, 811–12; Marie de Sévigné to Françoise-Marguerite de Grignan, 15 January 1690). The so-called doctrine of *attrition* was condemned on 24 September 1665 and 4 March 1679 by Popes Alexander VII and Innocent XI (Denzinger, H. J. D. (1957), 1101, 1155, 1156, 1157).
51. Belloc, H. J. P. R. (1920), 355.
52. There are reckoned to be 7 in Letter V, 16 in Letter VI, 19 in Letter VII, 28 in Letter VIII, 26 in Letter IX and 35 in Letter X (Belloc, H. J. P. R. (1920), 357 n. 1) although these numbers do not add up to 132.
53. Belloc, H. J. P. R. (1920), 372–3.
54. In Letter VII (731; *107*), (729–30; *105*), (731; *106*), (730; *105*), (734–5; *111*), (733; *109*), (733; *109*), (733; *109*).
55. These instructions were condemned on 24 September 1665 by Pope Alexander VII and on 10 November 1752 by Pope Benedict XIV (Denzinger, H. J. D. (1957), 1102, 1491).
56. Sánchez, T. (1615), II iii 6 (13) [1613].
57. Filliucci, V. (1633–4), XXV, xi (331).
58. Belloc, H. J. P. R. (1920), 373.
59. Cognet, L. (1965), 136 n. 2. A similar instruction was condemned on 24 September 1665 by Pope Alexander VII (Denzinger, H. J. D. (1957), 1126).
60. Tanner, A. (1621–7), III 1519.
61. Belloc, H. J. P. R. (1920), 373.
62. Belloc, H. J. P. R. (1920), 371.

63. Decrees of Popes Alexander VII and Innocent XI (Denzinger, H. J. D. (1957), 1108, 1117–18/1180, 1158, 1164, 1186).
64. Belloc, H. J. P. R. (1920), 373.
65. de Valencia, G. (1591), III 1042.
66. Lessius, L. (1606), II, xxi, 16 [1605].
67. This is borne out by a careful study of the footnotes to Cognet, L. (1965).
68. Kirk, K. E. (1927), 394.
69. Cognet, L. (1965), 103 n. 2.
70. Belloc, H. J. P. R. (1920), 355.

CHAPTER 6 THE CHRISTIAN LIFE

1. Pirot, G. (1657), 101, 107, 108, etc.
2. Pirot, G. (1657), 26, 28.
3. Pirot, G. (1657), 81.
4. Pirot, G. (1657), 123.
5. Pirot, G. (1657), 162–3.
6. Pirot, G. (1657), 87.
7. Pirot, G. (1657), 2, 176.
8. Pascal adapts 1 Corinthians IX 16.
9. Psalm CXIX 126–8.
10. The vicissitudes of the Jansenists' struggle against the Jesuits can be followed in *Annales* (1764–71), IV and V.
11. Matthew X, 34.
12. Ecclesiastes III, 8.
13. Matthew VIII, 25.
14. 2 Kings XIV, 17.
15. Luke XVI, 8.
16. The third and fourth statements in the name of the parish priests of Paris are not from Pascal's pen but, it would seem, from Arnauld's and Nicole's, respectively (**VII 353**); and very turgid they are.
17. This self-assessment has been preserved for posterity by Pascal's niece Marguerite Périer (**VII 353**).
18. Migne, J.-P. (1846), 69: 'quia præcidendæ unitatis nulla est justa necessitas' (*Contra Epistolam Parmeniani*, II, 11, 25).
19. Revelation IV 11.
20. *King Lear*, IV, 1.
21. Council of Trent (13 January 1547), Session VI, Chapter XIII.
22. Romans XII, 12.
23. Council of Trent (13 January 1547), Session VI, Chapter XI.
24. A parallel system of thought, and a neater solution of the dilemma, is to be found in Spinoza: 'I fully admit that God is not angry, and that all things come to pass according to His decision. . . He who is unable to control his desires, and to restrain them through fear of the laws, although he must be excused for his weakness, is nevertheless unable to enjoy peace of mind, and the knowledge and love of God, but

necessarily perishes' (Spinoza, B. (1928), 358; Spinoza to Henry Oldenburg, 7 February 1676).

25. John Dryden, *The Hind and the Panther*, 1687, I, 105.
26. There are also some grounds for believing that the *Prayer to God Concerning the Proper Use of Illnesses* may have been composed quite early in Pascal's lifetime: perhaps in 1647, when it seems that he was temporarily paralysed in the legs (**IX 321**).
27. Matthew XXVI 39.
28. *L'Annonce Faite à Marie*, III, 2.
29. In October–November 1660?
30. Matthew XIX, 16–22.
31. Luke XXI, 2–4.
32. Mark X, 25.
33. Nicole, P. (1670), 269–70.
34. Nicole, P. (1670), 270.
35. Pascal, seemingly adopting a form of tychism, here accepts the role of chance within a predetermined world.
36. Alexander Pope: *An Essay on Man, Epistle I*, line 294.
37. Matthew XXII, 21.
38. Romans XIII, 1.
39. Colossians III, 5, 1 Thessalonians IV 5.
40. 1 Corinthians XIII.
41. Hebrews XI, 13.

CHAPTER 7 HUMAN NATURE

1. Damien Mitton (1618–90) was a wealthy bourgeois, a worldly gambler and an agnostic whose capacity for doubt extended to almost every department of his life. He was a man of the utmost refinement and culture, and the author of 'Thoughts on *honnêteté*' and a 'Description of the *honnête homme*' both published in de Saint-Évremond, C.-M. de St-D. (1680), 1–7 and 8–12. A friend of Pascal from about 1651 onwards, Mitton is mentioned three times in the *Thoughts* (597*, 642*, 853*), always as the embodiment of smiling scepticism.
2. 1 John IV, 16.
3. de Chateaubriand, F.-R. (1802), III, 66.
4. Aristotle, *Poetics*, XI 2–5.
5. *Thought* 183*.
6. Notably from the 'Apologia of Raimond Sebond', where Montaigne even writes (de Montaigne, M. E. (1962), 563) of the 'crime' of crossing a river. This is echoed by Pascal in, for example, *Thoughts* 20* and 51*.
7. Pascal's father, uncle, grandfather and brother-in-law were all trained in the law.
8. Newman, J. H. (1870), 281–322: 'Informal Inference'.
9. Hammarskjöld, D. (1964), 136.
10. von Leibniz, G.W. (1874), II, 173; Leibniz to Sophia of Brunswick-Lüneburg, Electress of Hanover, 12 June 1700. See on this subject Jovy, E. (1932).

11. It occurs in *Thoughts* 242*, 427*, 781*.
12. Isaiah LIII, 3–12.
13. Périer, É. (1670), *Pensées*, ed. L. Lafuma, 1962, 17.
14. Filleau de La Chaise, J. (1672), 13. On various problems associated with the authenticity of Filleau's account, see McKenna, A. (1988). See also Périer, É. (1670), *Pensées*, ed. L. Lafuma, 1962, 1–17.
15. Newman, J.H. (1870), 484.
16. Leibniz, somewhat ambivalent in his attitude towards the *Thoughts*, also acknowledged that Pascal had been right to discard 'certain abstract metaphysical theories which will vanish into smoke and in which there is nothing of substance, nothing to move minds . . . : this, undeniably, is absolutely true' (von Leibniz, G.W. (1923–), II, i, 112; Leibniz to Johann Friedrich of Brunswick-Lüneburg, Duke of Hanover, 21 May 1671, Appendix, 'On the Resurrection of the Dead').

CHAPTER 8 THE WAGER

1. E.g., Migne, J-P. (1845), 235–41, Sermon XXXVIII; 442–4, Sermon LXX; etc. Whether or not an appreciation of the probability calculus is needed in order to undertake these risks is, of course, a matter for conjecture.
2. E.g., Nicole, Voltaire, Laplace, Renan and Claudel.
3. Kierkegaard, S.A. (1944) [1843]. But it was mainly between 1850 and 1854 that Kierkegaard was influenced by Pascal.
4. Psalm CXIX, 36.
5. Matthew XVIII, 20.

CHAPTER 9 THE SCRIPTURES

1. E.g., *Thought* 857*.
2. Newman, J.H. (1898–1901), VIII, 76–90: 'Miracles no Remedy for Unbelief'. Note page 77: 'Nothing is gained by miracles, nothing comes of miracles, as regards our religious views, principles, and habits.'
3. Malebranche, N. (1958–1970), XII, 177 and *passim*.
4. von Leibniz (1923–), II, i, 271–3: 'Quod ens perfectissimum existit.'
5. On Newton's view of God see Koyré, A. (1972), *passim*.
6. von Leibniz, G.W. (1923–), I, iii, 574; Leibniz to Veit Ludwig von Seckendorff, June 1683: 'Puto enim, Deum non tantum in historia sacra et civili aut etiam naturali nobis loqui, sed et intus in Mente nostra, per abstractas illas a materia æternasque Veritates.'
7. Kant, I. (1966), II, 559: 'Now I maintain that all attempts of reason to establish a theology by means of mere speculation are completely fruitless and, by virtue of their essential character, null and void; but also that the principles of reason as applied to nature do not in any way lead to a theology, and consequently that there can be no rational

theology whatever unless it is founded upon the laws of morality or uses these as a touchstone' [1781].

8. Hume, D. (1882), II, 88–108 ('Essay on Miracles').
9. Smart, R.N. (1964), 27–9.
10. See Brams, S.J. (1980). The probability calculus, so prominent a feature of human games and gambling, is perhaps absent from *this* ludic exercise because it has been so prominent a feature of the Wager: hence, for Pascal, the irrelevance of Hume's argument against miracles.
11. Étienne Périer confirms the all-importance of prophecy in his uncle's apologetical plan (Périer, É. (1670), *Pensées*, ed. L. Lafuma, 1962, 21).
12. E.g., *Thoughts* 318*, 236*, 255*, 257*, 263*.
13. John XVIII, 36.
14. Hegel, G.W.F. (1832), II, 156 [*circa* 1826].
15. E.g., de Montaigne, M.E. (1962), 415–589, 'Apologia of Raimond Sebond' [1580].
16. E.g., Isaiah LIII,, Jeremiah XXXI 15, Zechariah IX 9, Micah V 1–2.
17. E.g., Psalms II, XLVII, LXXII, XCIII,, CX.
18. Or *types*, in Newman's phrase: see his sermon 'Moses the Type of Christ' (Newman, J.H. (1898–1901), VII 118–32).
19. Isaiah LIII, 3–12. Cf. *Thought* 487*.
20. Psalm CXVIII, 22. Cf. *Thought* 487*.
21. Deuteronomy XVIII, 15.
22. John II 18–22, Mark XIII, 1–23.
23. Jeremiah XIII, 19, XX, 4, XXV, 11. Cf. *Thought* 349*. Newman has treated the same theme in one of the finest of his sermons, 'Omnipotence in Bonds' (Newman, J.H. (1900), 75–90), preached at Dublin on 11 January 1857.
24. Revelation XIII, 8.
25. Pascal actually suggests (746*), erroneously, that the ancient historians Josephus and Tacitus overlooked Jesus altogether.
26. Josephus, *Antiquities of the Jews*, XVIII, ch. 3 (3).
27. Pliny, *Letters*, X, letter 97 (Pliny to the Emperor Trajan, 112 A.D.).
28. Tacitus, *Annals*, XV, ch. 44.
29. Suetonius, *Lives of the Twelve Cæsars*, *Claudius*, ch. 25, *Nero*, ch. 16.
30. E.g., Numbers XXI, 8–9 (Moses's rod becomes a serpent), I Kings XVII, 17–24 (Elijah raises the widow's son from the dead), II Kings V, 1–14 (Elisha cures Naamon's leprosy)
31. The function of miracles within the *Thoughts* is, however, discounted in Hammond, N.G. (1992). Ernst, P. (1989), 175 even asserts (I translate): 'we find that the argument from miracles is irrevocably eliminated, vigorously discarded, once and for all, without any ambiguity whatsoever.'
32. Ernst, P. (1989), 166–9.
33. John IV, 48.
34. Ernst, P. (1989), 168.
35. E.g., Acts IX 40–42 (St Peter raises Tabitha from the dead), XIII, 6–12 (St Paul smites Bar-Jesus with blindness).
36. Wright, C.J. (1930), 60.

37. Locke, J. (1794), II, 239.
38. Butler, J. (1736), 236.
39. Berkeley, G. (1871), I, 198–9 [1710], III, 115–16 [1712].
40. Wright, C.J. (1930), 77.
41. 1 Corinthians XI, 23–5.
42. See also *Thoughts* 260*, 274*.
43. de Laplace, P.-S. (1814), 16–17.
44. Essentially, *Thoughts* 351*–76*.
45. Kant, I. (1966), II, 557: 'dem *Deisten* allen Glauben an Gott absprechen, und ihm lediglich die Behauptung eines Urwesens, oder obersten Ursache, übrig lassen' [1781].
46. *Thought* 1001* of the first Lafuma recension (page 355 of the Penguin Classics translation).
47. E.g., Voltaire (1974), 35; Voltaire to the Marquis de Villevielle, 26 August 1768.

CHAPTER 10 PASCAL, 'DEVOUT GEOMETRICIAN'

1. Bourget, P.-C.-J. (1922), I 211.
2. Huygens, C. (1888–1950), II, 8; Claude Mylon to Christiaan Huygens, 2 March 1657. The reason for this seclusion is clear. Pascal had been busily laying the foundations of his apologetical work, composing many *Thoughts* during the spring of 1657 (Ernst, P. (1989), 139, 166–9).
3. Huygens, C. (1888–1950), I, 493.
4. Huygens, C. (1888–1950), II, 494; Pierre de Carcavi to Christiaan Huygens, 28 September 1656.
5. Expressed as a decimal fraction: 1156.8314 to 1. This is known as the Gambler's Ruin problem, and even as Pascal's Problem. See Edwards, A. W. F. (1983). Edwards, A.W.F. (1987), 153–6 speculates on Pascal's way of solving this problem since he himself left no indication of the method employed.
6. This story is broadly corroborated by Leibniz in a letter to Thomas Burnett of Kemney dated 1 February 1697 (von Leibniz, G. W. (1875–90), III, 195–6). See also Mesnard, J. (1965), 648–9.
7. E.g., von Leibniz, G.W. (1874), II, 175; Leibniz to Sophia of Brunswick-Lüneburg, Electress of Hanover, 12 June 1700.
8. Leibniz discovered the differential calculus in 1675 (Aiton, E.J. (1985), 57).
9. Hofmann, J. E. (1974), 89–91; Aiton, E. J. (1985), 42, 49.
10. Humbert, P. (1947), 202. The details of the challenge are recounted in Hara, K. (1971b), 95–7.
11. Concerning Galileo's work on the cycloid see Hara, K. (1971a), 245, n. 27.
12. Concerning Mersenne's work on the cycloid see Hara, K. (1971a), 245, n. 28.
13. Torricelli, E. (1644).
14. Watson, J. D. (1968).
15. See **VIII 185, 183.**

16. Humbert, P. (1947), 204.
17. Humbert, P. (1947), 209.
18. Not, however, made public until Cavalieri, B. (1635).
19. Descartes, R. (1897–1909), VI 407–24 [1637].
20. The letter in which Pascal sets his prize questions on the cycloid was published early in June 1658 (**VIII 26 n. 1**); Sluse suspects that Pascal is the challenger in a letter dated 16 November 1658 (**VIII 227–9**); Huygens is convinced of the fact by 5 February 1659 (**IX 176**). Cf. **VIII 12–13** (letter from René de Sluse to Pascal, 6 July 1658).
21. Probably the *Fifth Statement by the Parish Priests of Paris.*
22. For Wallis's part in the cycloid competition see Scott, J. F. (1938), 154–5; Hara, K. (1969), Hara, K. (1971a), Hara, K. (1971b).
23. The first prize was, therefore, worth 400 francs (or approximately £2000 at today's values), whilst the second prize was equal to half that amount.
24. Humbert, P. (1947), 206.
25. This problem is based on one originally posed by Kepler in his *Supplement to the Stereometry of Archimedes* (Whiteside, D.T. (1960), 108–9). Wren's solution of it was published in Wallis, J. (1659), 80 and *Philosophical Transactions,* 17 November 1673, 6146–9. See also **VIII 142**; Oldenburg, H. (1965–86), X, 42; John Wallis to Henry Oldenburg, 23 June 1673; ibid., X, 279–80; John Wallis to Henry Oldenburg, 4 October 1673. Wren proved, contrary to Descartes, that two curves were capable of rectification.
26. A French version of the *Third Circular Letter Concerning the Cycloid* exists in MS form, dated 7 October 1658.
27. Humbert, P. (1947), 216.
28. Jean de Beaugrand was Fermat's Parisian correspondent, an influential amateur mathematician who was secretary to the Royal Household and editor of several of Viète's works.
29. Torricelli, E. (1644), II, 85.
30. See **VIII 182** (letter from John Wallis to Christiaan Huygens); Wallis, J. (1659), 77.
31. Bertrand, J.-L.-F. (1891), 324.
32. The facts appear to be these: that Galileo, not Mersenne, was the first mathematician to notice and study the cycloid; that Roberval, not Torricelli or Fermat, was the first, however, who actually squared the cycloid; that Torricelli, on the other hand, on the basis of Roberval's initial discovery, had gone off along his own line of research, producing fruitful results; and that, above all, Torricelli was entirely guiltless of the charge that he had plagiarized the papers communicated by Beaugrand to Galileo.
33. Cavalieri, B. (1635).
34. For an outline of the content of the *Letters from Dettonville to M. de Carcavi,* see Marie, M. (1883–8), IV 189–202.
35. Bourbaki, N. (1960), 207.
36. For an outline of the content of the 'Treatise on Right-Angled Trilaterals and their Ungulæ' (247–67) see Marie, M. (1883–8), IV 202–16. For a discussion of the 'Treatise on the Sines of Quarter-Circles'

(275–82) see *ibid.*, 216–26. Concerning the 'Treatise on Circular Arcs' (283–97) see *ibid.*, 226–9. Concerning the 'Short Treatise on Circular Solids' (298–303) see *ibid.*, 229–30. The excursus 'Properties of the Sums of Simple, Triangular and Pyramidal Series' (268–74) developed Pascal's earlier work *The Summing of the Powers of Integers*.

37. von Leibniz, G.W. (1849–60), III, 72–3 n.; Leibniz to Jakob Bernoulli, April 1703 (*deleted postscript*).
38. Boyer, C.B. (1959), 203.
39. Newton, I. (1959–77), II, 61, 68, 75; Leibniz to Henry Oldenburg, 17 August 1676.
40. But Wallis has a claim to priority in this field (Vacca, G. (1903), 3–4).
41. Huygens, C. (1888–1950), II, 397–402; January 1659.

CHAPTER 11 CONCLUSION

1. Newton, I. (1687), Book I; Aiton, E.J. (1972), 99–105. Leibniz also used mathematics in order to refute Cartesian physics (Aiton, E.J. (1985), 128, 245).
2. As Fermat increasingly devoted himself to number theory.
3. E.g., the statements about vortices in *The Principles of Philosophy*: Descartes, R. (1897–1909), VIII, 116–66, IX, 136–71 [1644].
4. Subject to the reservation that Pascal may not have carried out all his experiments, e.g., the experiment of the self-enclosed vacuum, which may have been conducted by Adrien Auzout. In his *Treatise on the Equilibrium of Liquids* and *Treatise on the Weight of the Air Mass* Pascal does not explicitly state that all the experiments he describes have actually been performed.
5. Hacking, I. (1973), *passim*.
6. That the earth and the other eight solar planets all rotate around the sun (1514). It is similar to Galileo's point of view (1632) in that both men were of the opinion that the earth moved.
7. Leibniz has made the same point: von Leibniz, G.W. (1923–), II, i, 302; Leibniz to Johann Friedrich of Brunswick-Lüneburg, Duke of Hanover, probably written in 1677.
8. de Montaigne, M.E. (1962), 553, 'Apologia of Raimond Sebond' [1580].
9. 10 August 1660.
10. See Humbert, P. (1947), 88.
11. Pascal to Étienne Noël, 29 October 1647. But ten years later, in the eighteenth *Provincial Letter*, Pascal seemed to have been converted to Galileo's view (900; 295–6).
12. Carré, J.-R. (1935), 34–7.
13. In so far as he created co-ordinate geometry in 1637.
14. Descartes, R. (1952), 42–5, 52, 61–9, 72–89, 111–12: *Rules for the Direction of the Mind* [written in 1628].
15. Belaval, Y. (1960), *passim*.
16. Descartes, R. (1952), 1013–5; Descartes to Jean-Baptiste Morin, 13 July 1638.

17. Descartes, R. (1952), 1013–5; Descartes to Jean-Baptiste Morin, 13 July 1638. Descartes, R. (1952), 1289; Descartes to Marin Mersenne, 13 December 1647.
18. The year of the first publication of the *Thoughts*.
19. The year of the first publication of *Thought* 1001* in Marguerite Périer's 'Mémoire sur la Vie de M. Pascal' (41).
20. In the Psalms.
21. In The Song of Solomon.
22. See Koyré, A. (1922), *passim*.
23. 'What sort of freak then is man! How novel, how monstrous, how chaotic, how paradoxical, how prodigious! Judge of all things, feeble earthworm, repository of truth, sink of doubt and error, glory and refuse of the universe!' (*Thought* 131*)
24. 'The last act is bloody, however fine the rest of the comedy may be. They throw soil over your head and it is all over for evermore' (*Thought* 165*).
25. 'The never-ending silence of those infinite spaces fills me with dread' (*Thought* 201*).
26. Daudet, A. (1931), 51 [1889].
27. Disraeli, B. (1832), II, 206.
28. Rousseau, J.-J. to Voltaire, 18 August 1756.
29. Pascal does not quite say love-making. Sexual intercourse is compared to sneezing in *Thought* 795* and is viewed as an act of weakness.
30. Broad, C. D. (1958), 32 has taken this view.
31. For the full extent of his critique of the Wager argument, see de Condorcet, M.-J.-A.-N. C. (1847–1849), IV, 513–14, *Notes on Voltaire* (1778).
32. Ore, O. (1960), 417.
33. Nicole, P. (1730–5), VIII, 242; Pierre Nicole to Marquis Renaud de Sévigné [*circa* 1672].
34. On determinism see also Hacking, I. (1990), *passim*.
35. Descartes, R. (1952), 588–9, *The Principles of Philosophy* [1644].
36. Spinoza, B.: Part V, 'Concerning the Power of the Intellect or Human Freedom' [1677].
37. 1 Peter I, 20.
38. In *On Heresies*, XLII and *On Baptism*, IV, XVII, XXIV. But St Augustine was also the author of *On Free Will*.
39. St Thomas Aquinas, 1923–9, III 163 [1258–9].
40. Hume, D. (1882), II, 28–65 [1748].
41. Hulme, T. E. (1924), 22. Hulme specifically refers to *Thought* 136*.
42. Quoted in Faber, G. C. (1933), 20.
43. Newman, J. H. (1870), 484.
44. 1 Peter II, 2.
45. de Montesquieu, C.-L. de S. (1949–51), I, 166, Letter XXIV [1721].
46. Tertullian, *Concerning the Flesh of Christ*, V: 'It is certain *because* it is impossible' (my italics).
47. Eliot, T. S. (1969), 105 [1928]. Cf. Newman, J.H. (1900), 75–90: 'Omnipotence in Bonds', a sermon preached in Dublin on 11 January 1857.

48. Now at Edinburgh, and painted between 1644 and 1648.
49. The veneration of the Mother of God was, however, confirmed by the Council of Ephesus in 431.
50. Not Pascal's successors Voltaire or Kant, however. Nor Samuel Johnson, who remarked to Boswell on 19 September 1777: 'If I had . . . no reference to futurity, I would spend my life in driving briskly in a post-chaise with a pretty woman' (Boswell, J. (1934–64), III, 162).
51. Nicole, P. (1730–5), VIII, 243; Pierre Nicole to Marquis Renaud de Sévigné [*circa* 1672].
52. Claudel, P. (1949), 184; Paul Claudel to André Gide, 7 December 1911.
53. Nicole, P. (1730–5), VIII, 242; Pierre Nicole to Marquis Renaud de Sévigné [*circa* 1672].
54. Few in Pascal's day had not been baptized, despite Mersenne's belief – which he later revised downwards – that Paris contained 50 000 atheists (Mersenne, M. (1623), 671): in the year of Pascal's birth the population of the capital was at least 415 000 (Mousnier, R. (1961), I, 21–2), and some people believed it was higher still! Pascal agreed with Mersenne that there were many agnostics and atheists ('impies et . . . athées') in Paris (43).
55. Atkins, P.W. (1992), 12.
56. Voltaire (1751), II, 171.
57. Concerning Pascal's use of language, see Davidson, H.M. (1965), Demorest, J.-J. (1957), Descotes, D. (1988), Jungo, M. (1950), Le Guern, M. (1969a), Le Guern, M. (1969b), Maggioni, M.J. (1950), Norman, B. (1973), Norman, B. (1988), Russell, O.W. (1977), Schärer, K. (1980) and Topliss, P.M. (1966).
58. de Chateaubriand, F.-R. (1802), III, 66.
59. de Goncourt, E. and J. (1956), I 1367 (21 December 1863).
60. Gibbon, E. (1907), 75: 'From the *Provincial Letters* of Pascal, which almost every year I have perused with new pleasure, I learned to manage the weapon of grave and temperate irony even on subjects of ecclesiastical solemnity' [*circa* 1790]. Cf. Prothero, R.E. (1896), II, 396, n. 4.
61. Pater, W.H. (1895), 178–9.
62. E.g., de Montaigne, M.E. (1962), 472, 'Apologia of Raimond Sebond' [1580]. See Bart, B.F. (1955).
63. Jovy, E. (1930).
64. St Bonaventura (1937), 60 [1259].
65. Vincent de Beauvais (1624), I, 20 and IV, 2 [*circa* 1260].
66. de Lorris, G. and de Meung, J. (1949), 322 [*circa* 1280].
67. Gerson, J. (1606), I, 366 [*circa* 1400].
68. Rabelais, F. (1573), *Tiers Livre*, 53.
69. de Montaigne, M.E. (1617); preface to the *Essays* by M. de J. de Gournay.
70. Voltaire (1785–9), LI, 84.
71. Apian, Stifel, Tartaglia, Stevin, Peletier, Oughtred, Hérigone. See Edwards, A.W.F. (1987), 50–6.
72. Archimedes, Apollonius of Perga, Pappus, Kepler, Mydorge, Cavalieri, Desargues, Roberval, Torricelli, Grégoire de Saint-Vincent.
73. Galileo, Torricelli, Descartes.

74. Stevin, Mersenne, Galileo, Torricelli.
75. Archimedes, Valerio, Cavalieri, Roberval, Guldin, Torricelli, Carcavi, Fermat, Grégoire de Saint-Vincent, Tacquet.
76. Koyré, A. (1956), 287: the words are René Taton's, in the discussion which followed Koyré's paper.
77. Newton, I. (1959–77), I, 416; Newton to Robert Hooke, 5 February 1676.
78. Turnbull, H.W. (1929), 80.
79. von Leibniz, G.W. (1923–), I, II, 112; II, i, 441; Leibniz to Johann Friedrich of Brunswick-Lüneburg, Duke of Hanover, probably written in 1678.
80. Cicero: *On Old Age*, XXIII [*circa* 45 B.C.]: 'Quod quidem ni ita se haberet, ut animi inmortales essent, haud optimi cuiusque animus maxime ad inmortalitatis gloriam niteretur' ('And, indeed, unless this were the case that souls are immortal, the souls of the noblest of men would not aspire above all things to an immortal glory').
81. Plato (1938), 137–8. Further proponents of a wager argument have included Arnobius (*circa* 300), al-Ghazali (*circa* 1100), Pomponazzi (1520), Cardano (1544), Ochino (1544), Hurtado de Mendoza (1617), Mersenne (1623, 1624), La Rivière (1626), Silhon (1634) and Sirmond (1635). After Pascal's death this tradition was continued by Craig (1699). The Wager is a form of the 'insurance-policy' argument, which (envisaged in much less severely mathematical terms than by Craig) has found favour with many throughout the ages even when they have not set it down in print.
82. To what extent, if at all, did Pascal *plan* this fragmentariness? N.G. Hammond (1992) stresses the point. Pascal himself writes in *Thought* 532*: 'I should be honouring my subject too much if I treated it in an orderly manner, since I am trying to show that it is incapable of that.' He thus presents his own *Thought* 696* in a new light.
83. Périer, É. (1670), *Pensées*, ed. L. Lafuma, 1962, 25.
84. Périer, É. (1670), *Pensées*, ed. L. Lafuma, 1962, 24.
85. Quoted in Hay, M.V. (1962), 7.
86. Renan, J.-E. (1926), 54–5; Renan to Ernest Havet, 6 July 1852.
87. Boileau, N. (1966), VII, 63.
88. Bayle, P. (1715), I, 47.
89. Berlin, I. (1953), 2.
90. Archilochus (1910), epode, fragment no. 103: 'this fox knows many things, that hedgehog there knows one really good one'.

Select Bibliography

EDITIONS OF PASCAL'S WORKS

The edition of Pascal's works from which quotations are principally taken is *Œuvres complètes de Blaise Pascal*, edited by J.-L.-A. Chevalier, published by Gallimard (Bibliothèque de la Pléiade), 1954. Page references to the Pléiade edition are printed in round brackets in the text of this book. Use has also been made of the masterly *Œuvres de Blaise Pascal publiées suivant l'ordre chronologique*, edited by L. Brunschvicg, P. Boutroux and F. Gazier, Édition des *Grands Écrivains de la France*, 11 vols, 1908–14. The text of this book contains many page references to the latter edition; they are in round brackets in bold type.

A dual system of referencing applies to quotations from *The Provincial Letters*. The first number, bracketed within the main text, refers to the Pléiade edition; this is provided at all times. The second (italicized) number indicates the context of those quotations in the Penguin Classics edition of *The Provincial Letters* (1967), a translation by A. J. Krailsheimer which, it has to be said, is unreliable.

The translations of all quotations, whether from French, Latin, Greek, German or Danish, are my own work. Book titles have likewise been translated into English whenever they are mentioned in the main text or endnotes; but the (Latin or French) titles of books published during or around Pascal's lifetime are printed in full in this list of references. There is only one reference number to each of the quotations from the *Thoughts*; these numbers are asterisked. They are thought numbers, not page numbers, and relate to the Penguin Classics edition, entitled *Pensées*, translated by A. J. Krailsheimer, 1966. The text of the Penguin Classics translation is that of the *Thoughts* edited by Louis Lafuma, originally published in 1951. This may be closer to Pascal's intended outline of his presumed apologia of the Christian religion than is the Pléiade edition.

Endnote references to quotations from multi volume works state, within brackets, the year of publication of the first and last volumes in the series: e.g., Huygens, C. (1888–1950), III refers to volume III of Huygens, C., *Œuvres complètes*, 22 vols published between 1888 and 1950; it is also stated in this bibliography when individual volumes of multi volume works were published, e.g., vol. III, 1890. Towards the end of Chapter 1 the reference within the body of the text occasionally supplies a date different from the reference in the endnotes: e.g., 'Benjamin Jowett, 1881' refers to *Sermons Biographical and Miscellaneous*, published in 1899, although the actual sermon on Pascal was preached eighteen years previously.

All dates within square brackets, whether in this bibliography or elsewhere, refer to the original date of composition or publication. Occasionally there is no endnote reference to an author (e.g., Boole, Rémond de Montmort, Soria-Butron) or nothing specific within an endnote concerning an author (e.g., Arnobius, Laymann, Valerio). Readers seeking clarification about such sources should turn straight to this bibliography, where in most cases the relevant works will be found listed in full.

Quotations from the Psalms refer to the numbering of the Psalms in the Church of England's Authorized Version of the Bible. This differs slightly from the method of numbering employed by the Roman Catholic Church.

Not all of the following works are actually cited in the text, but all are relevant to its subject-matter and all repay study. Sixty-one of them are prefixed with an asterisk signifying their particular importance. Louis Lafuma has also produced an excellent photographic edition of the *Thoughts*: *Le Manuscrit des Pensées de Pascal, 1662*, 1962.

OTHER PUBLISHED SOURCES

Abercrombie, N. J., *The Origins of Jansenism*, 1936.

Aiton, E. J., *The Vortex Theory of Planetary Motions*, 1972.

Aiton, E. J., *Leibniz, A Biography*, 1985.

Amico, F., *Cursus Theologicus*, 1642.

Amoudru, B., *Des* Pascalins *aux* pascalisants*: la Vie posthume des* Pensées, 1936.

Annales: *Annales de la Société des soi-disant jésuites*, 5 vols, 1764–71, vol. IV, 1769 [1648–58], vol. V, 1771 [1658–68].

Annat, F., *Le Libelle intitulé* Théologie morale des Jésuites, *contredit et convaincu en tous ses chefs*, 2nd edn, 1645.

Annat, F., *La Bonne Foi des jansénistes en la citation des auteurs reconnue dans les lettres que le secretaire du Port-Royal a fait courir depuis Pâques*, 1656.

Apian, P., *Eine neue und wohlgegründete Unterweisung aller Kaufmanns-Rechnung*, 1527.

Apollonius of Perga, *Conicorum Libri Quattuor*, 1566 [*circa* 225 B.C.].

Archilochus, *Supplementum Lyricum. Neue Bruchstücke*, 1910.

Archimedes, *Opera Nonnulla, Commentarii*, 2 parts, 1558.

Arnauld, A., *Lettre d'un docteur de Sorbonne à une personne de condition sur ce qui est arrivé depuis peu, dans une paroisse de Paris, à un seigneur de la Cour*, 1655.

Arnauld, A., *Seconde Lettre de Monsieur Arnauld . . . à un duc et pair de France pour servir de réponse à plusieurs écrits qui ont été publiés contre sa première lettre*, 1655.

Arnauld, A., *Nouveaux Éléments de Géométrie*, 1667.

Arnauld, A., *Œuvres*, 50 vols, 1775–83, vol. XXIX, 1779, pp. 74–94, *Théologie Morale des jésuites* [1643].

Arnauld, A. and Nicole, P., *La Logique, ou l'Art de Penser: contenant, outre les règles communes, plusieurs observations nouvelles propres à former le jugement*, 1662. (This is the so-called *Logique de Port-Royal* or *Port-Royal Logic*. The *Port-Royal Logic* is a collaborative work. Its joint authors are thought to be Antoine Arnauld (1612–94) and Pierre Nicole (1625–95), with crucial input from Pascal.)

Arnauld, A. and Nicole, P., *La Logique, ou l'Art de Penser: contenant, outre les règles communes, plusieurs observations nouvelles propres à former le jugement*, 2nd edn, 1664.

Arnobius, *Disputationum adversus Gentes, Libri VII* [*circa* 300 A.D.].

Aron, R., 'Note sur le Pari de Pascal', *Revue de Métaphysique et de morale*, January–March 1926, pp. 85–91.
Atkins, P. W., *The Creation*, 1981.
Atkins, P. W., 'The Nobility of Knowledge', *Daily Telegraph*, 11 May 1992.
St Augustine of Hippo, *De Libero Arbitrio* [*circa* 388, 395 A.D.].
St Augustine of Hippo, *Contra Epistulam Parmeniani* [*circa* 400 A.D.].
St Augustine of Hippo, *De Baptismo* [*circa* 401 A.D.].
St Augustine of Hippo, *De Hæresibus* [*circa* 429 A.D.].
St Augustine of Hippo, see Migne, J.-P. (1845) and (1846).
Azor, J., *Institutionum Moralium, in quibus Universæ Quæstiones ad Conscientiam recte aut prave Factorum pertinentes, breviter tractantur*, 3 vols, 1610–16.
*Baird, A. W. S., *Studies in Pascal's Ethics*, 1975.
Baldelli, N., *Disputationes ex Morali Theologia*, 1637.
Barker, J., *Strange Contrarieties: Pascal in England during the Age of Reason*, 1975.
Baron, M. E., *The Origins of the Infinitesimal Calculus*, 1969.
Barrès, A.-M., *L'Angoisse de Pascal. Édition suivie d'une étude sur les deux maisons de Pascal à Clermont-Ferrand*, 1918.
de Barry, P. B., *Le Paradis ouvert à Philagie, par cent dévotions à la Mère de Dieu. Aisées à pratiquer aux jours de ses Fêtes, & Octaves, qui se rencontrent à chaque mois de l'année*, 1655 [1636].
Bart, B. F., '*Abêtir* in Pascal and Montaigne', *Romance Philology*, August 1955, pp. 1–6.
de Barthélemy, É.-M., *Les Ducs et les duchés français avant et depuis 1789*, 1867.
Baudry de Saint-Gilles d'Asson, A., *Le Journal de M. de Saint-Gilles*, 1936 [1655–6].
Bauny, É., *Somme des Péchés qui se commettent en tous états, de leurs conditions et qualités*, 1646 [1634].
Bauny, É., *Theologia Moralis*, 1642.
Bayle, P., *Dictionnaire historique et critique*, 3rd edn, 3 vols, 1715, vol. III, pp. 138–45, 'Pascal (Blaise)' [1696].
Beitzinger, A.J., 'Pascal on Justice, Force, and Law', *Review of Politics*, April 1984, pp. 212–43.
Belaval, Y., *Leibniz critique de Descartes*, 1960.
*Belloc, H. J. P. R., 'An Analysis of the *Lettres provinciales*', *Studies*, September 1920, pp. 355–73.
Benkard, E., *Undying Faces. A Collection of Death Masks*, 1929.
Berkeley, G., *Works*, 4 vols, 1871.
Berlin, I., *The Hedgehog and the Fox*, 1953.
Bernoulli, J., *Ars conjectandi, opus posthumum. Accedit tractatus de seriebus infinitis, et epistola gallice scripta de Ludo Pilæ reticularis*, 1713.
Bertrand, J.-L.-F., *Blaise Pascal*, 1891.
Bevan, E. R., *Christianity*, 1932.
Bibliothèque Sainte-Geneviève, *Le Cabinet de Curiosités de la Bibliothèque Sainte-Geneviève, des origines à nos jours*, 1989.
de Bièvre, C., *Descartes et Pascal, examen critique de leur œuvre scientifique*, 1956.
Birch, T., *The History of the Royal Society*, 1757.

Bishop, M., 'Did Pascal die a Jansenist?', *Romanic Review*, 1939, pp. 352–60.

Blanchet, L., 'L'Attitude religieuse des Jésuites et les sources du pari de Pascal', *Revue de Métaphysique et de morale*, July–August, September–October 1919, pp. 477–516, 617–47.

Blondel, M., 'Le Jansénisme et l'anti-jansénisme de Pascal', *Revue de Métaphysique et de morale*, April–June 1923, pp. 129–63.

Boileau, J.-J., *Lettres de Monsieur B. sur différents sujets de morale et de piété*, 2 vols, 1737–42.

Boileau, N., *Œuvres complètes*, 1966.

St Bonaventura, *The Franciscan Vision*, 1937, incorporating *Itinerarium Mentis in Deum* [1259].

Boole, G., *The Mathematical Analysis of Logic, being an Essay towards a Calculus of Deductive Reasoning*, 1847.

*Bosmans, H., 'Sur l'Œuvre mathématique de Blaise Pascal', *Revue des Questions scientifiques*, 20 January 1924, pp. 130–60; 20 April 1924, pp. 424–51.

Boswell, J., *Life of Johnson*, ed. G. B. Hill and L. F. Powell, 6 vols, 1934–64, vol. III, 1934.

Bourbaki, N., *Eléments d'Histoire des mathématiques*, 1960, pp. 178–220, 'Calcul infinitésimal'. Bourbaki is a collective pseudonym adopted by a changing group of young French mathematicians.

Bourget, P.-C.-J., *Nouvelles Pages de critique et de doctrine*, 2 vols, 1922.

*Boyer, C. B., *The Concepts of the Calculus. A Critical and Historical Discussion of the Derivative and the Integral*, 1939.

Boyle, N., 'A Problem in the Interpretation of Pascal's Wager', *French Studies*, January 1980, pp. 20–44.

Boyle, R., *Hydrostatical Paradoxes, made out by New Experiments, for the most part physical and easy*, 1666.

Brams, S. J., *Biblical Games: A Strategic Analysis of Stories in the Old Testament*, 1980.

Bréhier, É., notice on R. Aron's article on Pascal [*see above*], *Revue d'Histoire de la philosophie*, January–March 1927, p. 122.

Bremond, H., *Histoire Littéraire du sentiment religieux en France depuis la fin des guerres de religion jusqu'à nos jours*, 12 vols, 1916–36, especially vol. IV (ch. 9), 1920.

Broad, C. D., *Personal Identity and Survival*, 1958.

*Broome, J. H., *Pascal*, 1965.

Brown, H., *Scientific Organizations in Seventeenth Century France, 1620–1680*, 1934.

Brunet, G., *Le Pari de Pascal*, 1956.

Brunet, G., *Un Prétendu traité de Pascal: le* Discours sur les passions de l'amour, 1959.

*Brunschvicg, L., *Descartes et Pascal, lecteurs de Montaigne*, 1944.

Busson, H., *La Pensée religieuse en France de Charron à Pascal*, 1933.

Butler, J., *The Analogy of Religion, Natural and Revealed, to the Constitution and Course of Nature*, 1736.

*Caillat, J., 'La Méthode scientifique selon Pascal. Formation et apprentissage de la méthode (1623–1653)', *Revue d'Histoire littéraire de la France*, April–June 1923, pp. 129–62, July–September 1923, pp. 273–99.

Caramuel de Lobkowitz, J., *Theologia Moralis Fundamentalis Decalogica*, 1652–3.
Cardano, G., *De Sapientia Libri Quinque, quibus Omnis Humanæ Vitæ Cursus Vivendique Ratio Explicatur*, 1544.
Cardano, G., *Liber de Ludo Aleæ*, translated by S.H. Gould, 1953 [*circa* 1560].
Cargile, J., 'Pascal's Wager', *Philosophy*, July 1966, pp. 250–57.
Carré, J.-R., *Réflexions sur l'Anti-Pascal de Voltaire*, 1935.
de Castro-Palao, F., *Opus Morale de Virtutibus et Vitiis*, 1631–47.
Cavalieri, B., *Lo Specchio Ustorio, overo trattato delle Settioni Coniche et alcuni loro Mirabili Effetti intorno al Lume, Caldo, Freddo, Suono e Moto ancora*, 1632.
Cavalieri, B., *Geometria Indivisibilibus Continuorum nova quadam Ratione Promota*, 1635.
Cavalieri, B., *Exercitationes Geometricæ Sex*, 1647.
Chaix-Ruy, J., *Le Jansénisme, Pascal et Port-Royal*, 1930.
Chamaillard, E., *Pascal Mondain et amoureux*, 1923.
Chambers, F.M., 'Pascal's Montaigne', *Publications of the Modern Language Association of America*, 1950, pp. 790–804.
de Chanteresne, see Nicole, P. (1670).
Charron, P., *De la Sagesse*, 1601.
Chasles, M., *Mémoire de Géométrie pure sur les propriétés générales des cônes du second degré*, 1830.
Chasles, M., *Mémoire de Géométrie sur les propriétés générales des coniques sphériques*, 1831.
Chasles, M., *Aperçu historique sur l'origine et le développement des méthodes en géométrie*, 1837.
de Chateaubriand, F.-R., *Génie du christianisme, ou Beautés de la Religion chrétienne*, 5 vols, 1802, vol. III, pp. 65–75, 'Suite des Moralistes'.
de Chénier, A.-M., *Œuvres inédites*, ed. A. Lefranc, 1914.
Chesneau, C., see Julien-Eymard d'Angers, (1954).
Chevalier, J.-L.-A., 'La Place du Pari dans l'Apologie de Pascal', *Mélanges de littérature, philologie et histoire offerts à M. Louis Arnould*, 1934, pp. 68–72.
Child, J.M., *The Early Mathematical Manuscripts of Leibniz*, 1920.
Church, A., 'Modern History of Logic', *Encyclopædia Britannica*, vol. XIV [1958], pp. 326–332.
Church, R.W., *Pascal and Other Sermons*, 1895.
Cicero, *De Senectute* [*circa* 45 B.C.].
*Clark, R., *Strangers and Sojourners at Port-Royal*, 1932.
Clark, W., *Pascal and the Port-Royalists*, 1902.
Claudel, P., *Paul Claudel et André Gide. Correspondance, 1899–1926*, 1949.
Claudel, P., *Œuvres en prose*, 1965.
Cognet, L., *Le Jansénisme*, 1961.
Cognet, L., 'Port-Royal tel que l'a connu Pascal', *Pascal et Port-Royal*, 1962, pp. 7–12.
Cognet, L., 'Pascal et Port-Royal', *Pascal. Textes du Tricentenaire*, 1963, pp. 98–104.
Cognet, L., *Les Provinciales*, ed. L. Cognet, 1965.
Coleman, F. X. J., *Neither Angel nor Beast: The Life and Work of Blaise Pascal*, 1986.
Collingwood, R.G., *The Idea of History*, 1946.
Collins, J., *God in Modern Philosophy*, 1960, pp. 327–40, 'Pascal on the Hidden God'.

Comte, A., *The Positivist Calendar of 558 Worthies of All Ages and Nations*, 1894 [drawn up in 1849].
de Condorcet, M.-J.-A.-N. C., *Esquisse d'un Tableau historique des progrès de l'esprit humain*, 1795 [1794].
de Condorcet, M.-J.-A.-N. C., Pensées *de Blaise Pascal. Notes de Voltaire et Condorcet*, 2 vols, 1803 [1776, 1778].
de Condorcet, M.-J.-A.-N. C., *Œuvres*, 12 vols, 1847–9, vol. IV, 1847, pp. 187–315, *Avertissements insérés par Condorcet dans l'édition complète des* Œuvres *de Voltaire* [1785], pp. 317–635, *Notes sur Voltaire* [1778].
Copleston, F.C., *A History of Philosophy*, 9 vols, 1946–75, vol. IV, 1958, pp. 153–73, 'Pascal'.
Couchoud, P.-L., 'L'Entretien de Pascal avec M. de Saci a-t-il eu lieu?', *Mercure de France*, 1 February 1951, pp. 216–28.
Council of Trent, *Canons and Decrees*, 1848.
Courcelle, P., *L'Entretien de Pascal et Sacy, ses sources et ses énigmes*, 1960.
Cousin, V., *Études sur Pascal*, 1857.
Cox, C.M., *The Early Mental Traits of Three Hundred Geniuses*, 1926.
Craig, J., *Theologiæ Christianæ Principia Mathematica*, 1699.
Croquette, B., *Pascal et Montaigne: Étude des réminiscences des* Essais *dans l'œuvre de Pascal*, 1974.
Cruickshank, J., *Pascal:* Pensées, 1983.
Cupitt, D., *Taking Leave of God*, 1980.
Dalton, P., 'Pascal's Wager: the Second Argument', *The Southern Journal of Philosophy*, 1975, pp. 31–46.
Dalton, P., 'Pascal's Wager: the First Argument', *International Journal for the Philosophy of Religion*, 1976, pp. 346–68.
Daudet, A., *La Doulou*, 1931 [1889].
David, F.N., *Games, Gods and Gambling. The Origins and History of Probability and Statistical Ideas from the Earliest Times to the Newtonian Era*, 1962.
David, F.N. and Barton, D.E., *Combinatorial Chance*, 1962.
Davidson, H.M., 'The Argument of Pascal's *Pari*', *Romanic Review*, April 1956, pp. 92–102.
*Davidson, H.M., *Audience, Words, and Art*, 1965.
Davidson, H.M., 'Pascal's Arts of Persuasion', *Renaissance Eloquence: Studies in the Theory and Practice of Renaissance Rhetoric*, 1983, pp. 292–300.
Davidson, H.M. and Dubé, P., *A Concordance to Pascal's* Pensées, 1975.
Davidson, H.M. and Dubé, P., *A Concordance to Pascal's* Les Provinciales, 2 vols, 1980.
Davies, P., *The Mind of God. Science and the Search for Ultimate Meaning*, 1992.
Dawkins, R., *The Blind Watchmaker*, 1988.
Delassault, G., *La Pensée janséniste en dehors de Pascal*, 1963.
Demahis, É., *La Pensée politique de Pascal. Le Milieu, la doctrine et son originalité*, 1931.
De Moivre, A., *The Doctrine of Chances: or, A Method of Calculating the Probability of Events in Play*, 1718.
De Moivre, A., *Annuities upon Lives; or, the Valuation of Annuities upon any Number of Lives; as also of Reversions*, 1725.
*Demorest, J.-J., *Pascal Écrivain, étude sur les variantes de ses écrits*, 1957.
De Morgan, A., *Formal Logic*, 1847.

Denzinger, H. J. D., *Enchiridion Symbolorum, Definitionum et Declarationum de Rebus Fidei et Morum,* 1955.
Denzinger, H. J. D., *The Sources of Catholic Dogma,* 1957.
Desargues, G., *Brouillon Projet d'une atteinte aux événements des rencontres d'un cône avec un plan,* 1639.
Descartes, R., *Œuvres,* 13 vols, 1897–1913: V, 1903, correspondence, May 1647–February 1650; VI, 1902, *La Géométrie* [1637]; VIII, 1905, *Principia philosophiæ* [1644]; translated as IX, 1904, *Principes de la philosophie;* X, 1908, *Regulæ ad directionem ingenii* [*circa* 1628, posthumously published 1701].
Descartes, R., *Œuvres et Lettres,* 1952.
Descotes, D., 'Pascal, Rhétoricien de la géométrie', *Cahiers de l'Association internationale des études françaises,* May 1988, pp. 251–71.
Descrains, J., *Jean-Pierre Camus (1584–1652) et ses* Diversités *(1609–1618), ou la culture d'un évêque humaniste,* 1985.
Desgrippes, G., 'Exégèses pascaliennes: le Pari', *Revue de philosophie,* 1935, pp. 313–32.
Desjardins, P., *La Méthode des classiques français. Corneille, Poussin, Pascal,* 1904.
Desmolets, P.-N., *Continuation des Mémoires de Littérature et d'histoire,* 11 vols, 1726–31, vol. V, part II, 1728, pp. 237–70, 'Entretien de M. Pascal et de M. de Sacy, sur la lecture d'Épictète et de Montaigne' [by N. Fontaine, precise date unknown].
Diana, A., *Resolutionum Moralium Partes Duodecim,* 9 vols plus Index, 1698 [1629–41].
Diderot, D. and d'Alembert, J. Le R., *Encyclopédie, ou Dictionnaire raisonné des sciences, des arts et des métiers,* ed. D. Diderot and J. Le R. d'Alembert, 17 vols, 1751–65, vol. I, 1751, pp. 680–84, 'Arithmétique (Machine)' [this article is by d'Alembert].
Dijksterhuis, E. J., *The Mechanization of the World Picture,* 1961, pp. 444–55, 'Pneumatics: Blaise Pascal'.
Disraeli, B., *Contarini Fleming. A Psychological Autobiography,* 4 vols, 1832.
Disraeli, B., *Coningsby; or, The New Generation,* 3 vols, 1844.
Dorival, B., 'L'Iconographie de Pascal au XVIIe siècle', *Pascal et Port-Royal,* 1962, pp. 91–8.
*Dorival, B., *Album Pascal,* ed. B. Dorival, 1978.
Ducas, A., *Le Pari de Pascal. Les Sources possibles de l'argument du Pari,* 1951.
Ducas, A., *Le* Discours sur les Passions de l'Amour *de Pascal,* 1953.
*Duchêne, R., *L'Imposture littéraire dans les 'Provinciales' de Pascal,* 1985.
Duclou, M., *Les Carrosses à cinq sols,* 1950.
Dugas, L. and Riquier, C., 'Le Pari de Pascal', *Revue Philosophique de la France et de l'étranger,* September 1900, pp. 225–45.
Dugas, R., *La Mécanique au XVIIe siècle,* 1950.
Duhamel, G., *Les Confessions sans pénitence,* 1941.
Eastwood, D. M., *The Revival of Pascal. A Study of his Relation to Modern French Thought,* 1936.
Edwards, A. W. F., 'Pascal and the Problem of Points', *International Statistical Review,* December 1982, pp. 259–66.
Edwards, A. W. F., 'Pascal's Problem: The *Gambler's Ruin*', *International Statistical Review,* April 1983, pp. 73–9.
*Edwards, A. W. F., *Pascal's Arithmetical Triangle,* 1987.

Eliot, G., *Middlemarch. A Study of Provincial Life*, 4 vols, 1871–72.
Eliot, T. S., 'The *Pensées* of Pascal', introduction to the Everyman's Library edition of *Pascal's Pensées*, 1931.
Eliot, T. S., *Complete Poems and Plays*, 1969.
*Ernst, P. , *Approches pascaliennes. L'Unité et le mouvement, le sens et la fonction de chacune des vingt-sept liasses titrées*, 1970.
*Ernst, P. , 'Géologie et Stratigraphie des *Pensées* de Pascal', doctoral thesis, University of Paris IV, 4 vols, 1989.
de Escobar y Mendoza, A., *Universæ Theologiæ Moralis Receptiores absque lite Sententiæ necnon Problematicæ Disquisitiones*, 1652.
de Escobar y Mendoza, A., *Liber Theologiæ Moralis, Viginti Quatuor Societatis Jesu Doctoribus Reseratus*, 1659 [1644].
Eusebius, *Historia ecclesiastica* [*circa* 328 A.D.].
Faber, G. C., *Oxford Apostles. A Character Study of the Oxford Movement*, 1933.
Faugère, A.-P., *Pensées, Fragments et lettres de Blaise Pascal*, 2 vols, 1844.
Faugère, A.-P., *Lettres, Opuscules et mémoires de Madame Périer et de Jacqueline, sœurs de Pascal, et de Marguerite Périer*, 1845.
de Fermat, P., *Œuvres*, 5 vols, 1891–1922.
Ferreyrolles, G., 'Pascal et la rédemption de la nature', *Méthodes chez Pascal*, 1979, pp. 285–95.
Filleau de La Chaise, J., *Discours sur les* Pensées *de M. Pascal, où l'on essaye de faire voir quel était son dessein . . .* , 1672.
Filliucci, V., *Quæstionum Moralium de Christianis Officiis in Casibus Conscientiæ*, 1633–4.
*Finch, D., *La Critique philosophique de Pascal au XVIIIe siècle*, 1940.
Fléchier, V.-E., *Mémoires sur les Grands-Jours tenus à Clermont en 1665–1666*, 1844 [1667].
Fletcher, F. T. H., *Pascal and the Mystical Tradition*, 1954.
Flew, A., 'Is Pascal's Wager the Only Safe Bet?', *Rationalist Annual*, 1960, pp. 21–5.
Fontaine, N., see Desmolets, P. -N. (1726–31).
Forton, J. [Sieur de Saint-Ange], *La Conduite du jugement naturel, où tous les bons esprits de l'un et l'autre sexe pourront facilement puiser la pureté de la science*, 3 vols, 1637–45.
*Francis, R., *Les* Pensées *de Pascal en France, de 1842 à 1942. Essai d'étude historique et critique*, 1959.
François, J., *L'Arithmétique, ou l'Art de compter toute sorte de nombres avec la plume et les jetons*, 1653.
Fric, R., *Les Pascal en Basse-Auvergne*, 1923. (Includes, pp. 37–41, 'Les Machines arithmétiques de Pascal'.)
Friedenthal, R., *Entdecker des Ich: Montaigne, Pascal, Diderot*, 1969.
Fry, T. C., *Probability and its Engineering Uses*, 1928.
Gazier, A., *Port-Royal au XVIIe siècle. Images et Portraits, avec des notes historiques et iconographiques*, 1909.
Gazier, A., *Blaise Pascal et Antoine Escobar, étude historique et critique*, 1912.
Gerson, J., *Opera*, 4 vols, 1606 [*circa* 1400].
Gibbon, E., *Autobiography*, 1907 [*circa* 1790].
Gide, A., *Journal, 1889–1939*, 1951.

Goldmann, L., *Le Dieu caché. Étude sur la vision tragique dans les* Pensées *de Pascal et dans le théâtre de Racine*, 1955.
Goldstine, H. H., *The Computer from Pascal to Von Neumann*, 1972.
de Goncourt, E. and J., *Journal. Mémoires de la Vie littéraire*, 4 vols, 1956.
Gouhier, H., 'Le *Mémorial* est-il un texte mystique?', *Blaise Pascal. L'Homme et l'Œuvre*, 1956, pp. 296–320.
Gouhier, H., 'Le *Pari* de Pascal', *Pascal e Nietzsche*, 1962, pp. 13–39.
Gouhier, H., 'Le Pari de Pascal', *Blaise Pascal, commentaires*, 1966, pp. 245–306.
Gouhier, H., *Blaise Pascal: Conversion et apologétique*, 1986.
*Gounelle, A., *La Bible selon Pascal*, 1970.
Green, J., *Journal, 1946–1950. Le Revenant*, 1951.
Greenstreet, W.J., *Isaac Newton, 1642–1727*, ed. W.J. Greenstreet, 1927.
Grégoire de Saint-Vincent, *Opus Geometricum Quadraturæ Circuli et Sectionum Coni*, 1647.
Grote, J., 'Pascal and Montaigne', *Contemporary Review*, July 1877, pp. 285–96.
Grubbs, H. A., *Damien Mitton, 1618–1690, bourgeois honnête homme*, 1932.
Guitton, J., *Pascal et Leibniz. Étude sur deux types de penseurs*, 1951.
Guldin, P., *De Centro Gravitatis Trium Specierum Quantitatis Continuæ*, 4 parts, 1635–41.
Hacking, I., 'Leibniz and Descartes: Proof and Eternal Truths', *Proceedings of the British Academy*, 1973, pp. 3–16.
*Hacking, I., *The Emergence of Probability. A Philosophical Study of Early Ideas about Probability, Induction and Statistical Inference*, 1975. (Includes, pp. 63–72, 'The Great Decision' [on the Wager].)
Hacking, I., *The Taming of Chance*, 1990.
Hammarskjöld, D., *Markings*, 1964.
*Hammond, N.G., 'Playing with Truth: Language and the Human Condition in Pascal's *Pensées*', D.Phil. thesis, University of Oxford, 1992.
Hara, K., 'Pascal et l'induction mathématique', *Revue d'Histoire des sciences et de leurs applications*, July–December 1962, pp. 287–302.
Hara, K., 'Pascal et Wallis au sujet de la cycloïde', *Annals of the Japan Association for Philosophy of Science*, 1969, pp. 36–57.
Hara, K., 'Pascal et Wallis au sujet de la cycloïde (II)', *Gallia*, 1971, pp. 231–49 (1971a).
Hara, K., 'Pascal et Wallis au sujet de la cycloïde (III)', *Japanese Studies in the History of Science*, 1971, pp. 95–112 (1971b).
Harrington, T.M., *Vérité et Méthode dans les* Pensées *de Pascal*, 1972.
Havet, E., *Pensées*, ed. E. Havet, 1881.
*Havet, E., *Pensées*, ed. E. Havet, 1899.
Hawking, S. W., *A Brief History of Time*, 1988.
Hay, M. V., *The Prejudices of Pascal concerning in particular the Jesuit Order and the Jewish People*, 1962.
Hegel, G. W. F., *Vorlesungen über die Philosophie der Religion*, 2 vols, 1832 [*circa* 1826].
Hérigone, P., *Cursus Mathematicus, Nova, Brevi et Clara Methodo Demonstratus*, 6 vols, 1634–42.
*Høffding, H., 'Pascal et Kierkegaard', *Revue de Métaphysique et de Morale*, April–June 1923, pp. 221–46.

*Hofmann, J. E., *Leibniz in Paris, 1672–1676. His Growth to Mathematical Maturity*, 1974.
Hollis, M. C., *The Jesuits. A History*, 1968.
*Howells, B. P., 'The Interpretation of Pascal's *Pari*', *Modern Language Review*, January 1984, pp. 45–63.
*Hubert, M.-L., *Pascal's Unfinished Apology. A Study of his Plan*, 1952.
Hulme, T. E., *Speculations. Essays on Humanism and the Philosophy of Art*, 1924.
*Humbert, P., *L'Œuvre scientifique de Blaise Pascal*, 1947.
Hume, D., *Philosophical Works. Essays Moral, Political and Literary*, ed. T. H. Green and T. H. Grose, 2 vols, 1882: II, 3–135, *An Enquiry Concerning Human Understanding* [1748]; II, 289–305, 'A Dialogue' [1750].
Hurtado, G., *Tractatus de Sacramentis et Censuris*, 2nd edn, 1633.
Hurtado de Mendoza, P., *Disputationes de Universa Philosophia*, 1617.
Hurtado de Mendoza, P., *De Tribus Virtutibus Theologicis*, 1631.
Hurter, H., *Nomenclator Literarius Recentioris Theologiæ Catholicæ Theologos Exhibens qui inde a Concilio Tridentino floruerunt Ætate, Natione, Disciplinis distinctos*, 3 vols, 1871–86.
Huxley, A. L., *Do What You Will*, 1929.
Huygens, C., *Œuvres complètes*, 22 vols, 1888–1950, vol. I, 1888; vol. II, 1889; vol. III, 1890; vol. XIV, 1920, *De Ratiociniis in Aleæ Ludo* (*Du Calcul dans les jeux de hasard*) [1656–7].
James, E. D., *Pierre Nicole, Jansenist and Humanist. A Study of his Thought*, 1972.
James, W., *The Varieties of Religious Experience. A Study in Human Nature*, 1902.
Jansen, C., *Augustinus, seu Doctrina S. Augustini de Humanæ Naturæ Sanitate, Ægritudine, Medicina adversus Pelagianos et Massilienses*, 1640.
*Jerphagnon, L., *Le Caractère de Pascal. Essai de caractérologie littéraire*, 1962.
Jordan, H. R., *Blaise Pascal: a Study in Religious Psychology*, 1909.
Journet, C., *Vérité de Pascal. Essai sur la valeur apologétique des* Pensées, 1951.
Jovy, E., 'Pascal et Saint-Ange', *Études pascaliennes*, vol. I, ed. E. Jovy, 1927.
Jovy, E., 'Pascal et la Tour Saint-Jacques (1648). A propos des expériences parisiennes de Pascal pour vérifier l'expérience du Puy de Dôme', *Études pascaliennes*, vol. III, ed. E. Jovy, 1928, pp. 7–22.
Jovy, E., 'La *Sphère infinie* de Pascal', *Études pascaliennes*, vol. VII, ed. E. Jovy, 1930, pp. 7–58.
Jovy, E., 'Les Antécédents de l'*infiniment petit* dans Pascal', *Études pascaliennes*, vol. VIII, ed. E. Jovy, 1932, pp. 9–77.
Jowett, B., *Sermons Biographical and Miscellaneous*, 1899.
Julien-Eymard d'Angers [the name in religion of C. Chesneau], *L'Apologétique en France de 1580 à 1670. Pascal et ses précurseurs*, 1954.
*Jungo, M., *Le Vocabulaire de Pascal étudié dans les fragments pour une Apologie*, 1950.
Kant, I., *Werke*, 6 vols, 1966, vol. II, *Kritik der reinen Vernunft* (*Critique of Pure Reason*) [1781].
Kepler, J., *Stereometria Doliorum Vinariorum . . . Accessit Stereometriæ Archimedeæ Supplementum*, 1615.
Keynes, J. M., *A Treatise on Probability*, 1921.
Kierkegaard, S. A., *Papirer*, 11 vols, 1900–48, vol. X, part 5, 1932.
Kierkegaard, S. A., *Either/Or*, 2 vols, 1944 [1843].
*Kirk, K. E., *Conscience and its Problems. An Introduction to Casuistry*, 1927.

Knowles Middleton, W.E., *The History of the Barometer*, 1964.
Knox, R. A., *Enthusiasm. A Chapter in the History of Religion with special reference to the XVII and XVIII Centuries*, 1950.
van Konijnenberg, J.T., *Courant pascalien et courant anti-pascalien de 1670 à 1734*, 1932.
*Koyré, A., 'Pascal savant', *Blaise Pascal, l'Homme et l'Œuvre*, 1956, pp. 259–85, plus 'Discussion', pp. 286–95.
Koyré, A., 'Leibniz and Newton', *Leibniz: A Collection of Critical Essays*, ed. H.G. Frankfurt, 1972, pp. 239–79.
*Krailsheimer, A. J., *Pascal*, 1980.
Kuhn, T. S., *The Structure of Scientific Revolutions*, 2nd edn, 1970.
Laberthonnière, L., *Essais de Philosophie religieuse*, 1903, pp. 193–229, 'L'Apologétique et la Méthode de Pascal'.
La Charité, R. C., 'Pascal's Ambivalence Toward Montaigne', *Studies in Philology*, April 1973, pp. 187–98.
Lachelier, J., 'Notes sur le Pari de Pascal', *Revue philosophique de la France et de l'étranger*, June 1901, pp. 625–39.
Lacombe, R.-E., *L'Apologétique de Pascal*, 1958.
Lafuma, L., *L'Auteur présumé du* Discours sur les Passions de l'amour, *Charles-Paul d'Escoubleau, Marquis d'Alluye et de Soudis*, 1950.
Lafuma, L., *Histoire des* Pensées *de Pascal (1656–1952)*, 1954.
Lafuma, L., *Le Manuscrit des* Pensées *de Pascal, 1662*, 1962.
Lagarde, P. -H.-J., 'The Scientific Method of Pascal', B.Litt. thesis, University of Oxford, 1933.
de La Hire, P., *Nouvelle Méthode en géométrie pour les sections des superficies coniques et cylindriques*, 1673.
de La Hire, P., *Nouveaux Éléments des sections coniques, les lieux géométriques, la construction ou effection des équations*, 1679.
de La Loubère, A., *Veterum Geometria Promota in septem de Cycloïde Libris et in duabus Adjectis Appendicibus*, 1660.
Lancelot, C. and Arnauld, A., *Grammaire générale et raisonnée*, 1660. (This is the so-called *Grammaire de Port-Royal* or *Port-Royal Grammar*. Its principal author is thought to be Claude Lancelot (1615–95) but Antoine Arnauld seems to have played a considerable part in its composition.)
Landsberg, P. T., 'Gambling on God', *Mind*, January 1971, pp. 100–104.
de Laplace, P.-S., *Essai philosophique sur les probabilités*, 1814.
Laporte, J., *La Doctrine du Port-Royal*, 1923.
Laporte, J., *Le Cœur et la raison selon Pascal*, 1950.
de La Rivière, P., *Angélique. Des Excellences et perfections immortelles de l'âme*, 1626.
Larroutis, M., 'Le Pari', *La Table ronde*, April 1962, pp. 147–9.
Laymann, P., *Theologia Moralis in Quinque Libros partita*, 1625.
Leavenworth, I., *A Methodological Analysis of the Physics of Pascal*, 1930.
*Le Guern, M., *L'Image dans l'Œuvre de Pascal*, 1969a.
Le Guern, M., 'Tendances constantes du langage de Pascal', *Le Français moderne*, January 1969, pp. 41–8 (1969b).
Le Guern, M., *Pascal et Descartes*, 1971.
Le Guern, M. and M.-R., *Les* Pensées *de Pascal: de l'anthropologie à la théologie*, 1972.

von Leibniz, G. W., *Ars Combinatoria*, 1690 [1666].
von Leibniz, G. W., *Mathematische Schriften*, 6 vols, 1849–60, vol. III, 1855.
von Leibniz, G. W., *Correspondance avec l'Électrice Sophie de Brunswick-Lunebourg*, 3 vols, 1874.
von Leibniz, G. W., *Philosophische Schriften*, 7 vols, 1875–1890, vol. III, 1887.
von Leibniz, G. W., *Sämtliche Schriften und Briefe*, 28 vols, 1923–, series I, vol. II, 1927; series I, vol. III, 1938; series II, vol. I, 1926.
Le Moyne, P., *Les Peintures morales. De la Doctrine des Passions, où il est traité de leur nature, et de leur modération; et les plus belles matières de la morale chrétienne sont expliquées*, 2nd edn, 1645 [1640–43].
Le Moyne, P., *La Dévotion aisée*, 1652.
Le Roy, G., *Pascal, savant et croyant*, 1957.
Lessius, L., *De Justitia et Jure cæterisque Virtutibus Cardinalibus*, 1606 [1605].
Levi, A. H. T., *French Moralists*, 1964.
von Leyden, W., *Seventeenth-Century Metaphysics: An Examination of Some Main Concepts and Theories*, 1971.
*Lhermet, J., *Pascal et la Bible*, 1931.
Ligonnière, R., *Préhistoire et Histoire des ordinateurs*, 1987.
Locke, J., *Works*, 9 vols, 1794.
Loeffel, H., *Blaise Pascal, 1623–1662*, 1987.
Lønning, P., *Cet Effrayant pari. Une 'Pensée' pascalienne et ses critiques*, 1980.
de Lorris, G. and de Meung, J., *Le Roman de la rose*, 5th edn, 1949 [*circa* 1236, *circa* 1280].
de Lugo, J., *Disputationes Scholasticæ et Morales*, 1644.
Macaulay, T. B., review of *Memoirs of the Life and Administration of the Right Honourable William Cecil, Lord Burghley . . .* [by E. Nares], *Edinburgh Review*, April 1832, pp. 271–96.
McKenna, A., 'L'Argument *Infini-Rien*', *Méthodes chez Pascal*, 1979, pp. 497–508.
McKenna, A., 'Filleau de La Chaise et la réception des *Pensées*', *Cahiers de l'Association internationale des études françaises*, May 1988, pp. 297–314.
McKenna, A., *De Pascal à Voltaire. Le Rôle des* Pensées *de Pascal dans l'histoire des idées entre 1670 et 1734*, 2 vols, 1990.
Magendie, M., *La Politesse mondaine et les théories de l'honnêteté en France au XVIIe siècle, de 1600 à 1660*, 2 vols, 1925.
*Maggioni, M. J., *The* Pensées *of Pascal: a Study in Baroque Style*, 1950.
Magnard, P., *Nature et Histoire dans l'apologétique de Pascal*, 2nd edn, 1980.
Mahoney, M. S., *The Mathematical Career of Pierre de Fermat, 1601–65*, 1973.
Maire, A., *Bibliographie de l'Œuvre scientifique de Blaise Pascal*, 1912.
Maire, A., *Bibliographie générale des Œuvres de Blaise Pascal*, 5 vols, 1925–7, vol. II, 1925; vol. V, pp. 84–8, 'Les Portraits de Pascal'.
de Maistre, J.-M., *Les Soirées de Saint-Pétersbourg, ou Entretiens sur le gouvernement temporel de la Providence*, 2 vols, 1821.
Maistrov, L. E., *Probability Theory: A Historical Sketch*, 1974.
Malebranche, N., *Œuvres complètes*, 21 vols, 1958–70, vol. XII, 1965, pp. 29–354, *Entretiens sur la Métaphysique et sur la religion* [1688].
Mandrou, R., *From Humanism to Science, 1480–1700*, 1978.
Mantoy, J., *Des* Pensées *de Pascal à l'*Apologie, 1955.
Marie, M., *Histoire des sciences mathématiques et physiques*, 12 vols, 1883–8, vol. IV, 1884, pp. 182–230, 'Blaise Pascal'.

*Marin, L., *La Critique du discours: sur la* Logique de Port-Royal *et les* Pensées *de Pascal*, 1975a.
*Marin, L., '"Pascal": Text, Author, Discourse . . .', *Yale French Studies*, May 1975, pp. 129–51 (1975b).
Maritain, J., 'Pascal apologiste', *Pascal. La Revue hebdomadaire*, 1923, pp. 56–72.
Maritain, J., 'La Politique de Pascal', *Revue universelle*, 1 August 1923, pp. 257–68.
Martini, R., *Pugio Fidei . . . adversus Mauros et Judæos*, 1651 [*circa* 1268].
Maurras, C.-M.-P., *Pascal puni. Conte infernal*, 1953.
Maynard, M.-U., *Les* Lettres provinciales . . . *et leur réfutation*, ed. M.-U. Maynard, 2 vols, 1851.
Maynard, T., *St Ignatius and the Jesuits*, 1956.
*Melzer, S. E., *Discourses of the Fall: a Study of Pascal's* Pensées, 1986.
de Méré, A. G., *Œuvres posthumes*, 1700, pp. 1–95, *De la Vraie honnêteté*.
Mersenne, M., *Quæstiones Celeberrimæ in Genesim*, 1623.
Mersenne, M., *L'Impiété des déistes et des plus subtils libertins découverte et réfutée par raisons de théologie et de philosophie*, 2 vols, 1624.
Mesnard, J., 'Les Séjours de Pascal à Port-Royal-des-Champs', *Société des Amis de Port-Royal*, 1951, pp. 15–28.
Mesnard, J., 'A la Recherche de la bibliothèque de Pascal', *Société des Amis de Port-Royal*, 1952, pp. 33–46 (1952a).
Mesnard, J., *Pascal, his Life and Works*, 1952b (preface by R.A. Knox).
*Mesnard, J., 'Introduction à l'Étude de Pascal mondain', *Annales Universitatis Saraviensis*, 1954, pp. 75–94.
Mesnard, J., 'Pascal à l'Académie Le Pailleur', *Revue d'Histoire des sciences et de leurs applications*, January-March 1963, pp. 1–10.
*Mesnard, J., *Pascal et les Roannez*, 1965.
*Mesnard, J., *Les* Pensées *de Pascal*, 1976.
de Meyer, A., *Les Premières controverses jansénistes en France*, 1917.
*Miel, J., *Pascal and Theology*, 1969a.
*Miel, J., 'Pascal, Port-Royal, and Cartesian Linguistics', *Journal of the History of Ideas*, April–June 1969, pp. 261–71 (1969b).
Migne, J.-P., *Patrologiæ Cursus Completus*, ed. J.-P. Migne, vol. XXXVIII, 1845, S. Aurelii Augustini, vol. V i.
Migne, J.-P., *Patrologiæ Cursus Completus*, ed. J.-P. Migne, vol. XLIII, 1846, S. Aurelii Augustini, vol. IX.
Mill, J. S., *Autobiography*, 1873.
Mitton, D., *see* de Saint-Évremond, C.-M. de St-D. (1680).
Moles, E., 'Pascal's Theory of the Heart', *Modern Language Notes*, May 1969, pp. 548–64.
Moles, E., 'Logic and Singularity in Pascal's Wager', *French Studies*, October 1982, pp. 397–409.
Molina, L., *Liberi Arbitrii cum Gratiæ Donis, Divina Præscientia, Providentia, Prædestinatione, et Reprobatione Concordia*, 1588.
*Molino, J., 'La Raison des effets', *Méthodes chez Pascal*, 1979, pp. 477–96.
Monod, A., *De Pascal à Chateaubriand: les Défenseurs français du christianisme de 1670 à 1802*, 1916.
de Montaigne, M. E., *Essais*, prefaced by M. de J. de Gournay, 1617.
de Montaigne, M. E., *Œuvres complètes*, 1962, pp. 11–1097, *Essais* (pp. 415–589, 'Apologie de Raimond Sebond' [1580]).

de Montesquieu, C.-L. de S., *Œuvres complètes*, 2 vols, 1949–51, vol. I, pp. 129–386, *Lettres persanes* [1721].
*Morgan, J.S., 'Pascal's Three Orders', *Modern Language Review*, October 1978, pp. 755–66.
Morot-Sir, É., *La Métaphysique de Pascal*, 1973.
Morot-Sir, É., 'Du Nouveau sur Pascal?', *Romance Notes*, winter 1977, pp. 272–9.
Mortimer, E., *Blaise Pascal. The Life and Work of a Realist*, 1959.
Mousnier, R., *Paris au XVIIe siècle*, 3 vols, 1961.
Mury, G., *Un Marxiste peut-il comprendre Pascal?*, 1961.
Mydorge, C., *Prodromi Catoptricorum et Dioptricorum, sive Conicorum operis . . . Libri Primus et Secundus*, 1631.
Mydorge, C., *De Sectionibus Conicis*, 1644.
Nadler, S.M., *Arnauld and the Cartesian Philosophy of Ideas*, 1989.
Nedeljkovic, D., *La Pensée philosophique créatrice de Pascal, avec une note sur la conception pascalienne de l'amour*, 1925.
Nelson, R.J., *Pascal: Adversary and Advocate*, 1981.
Newman, J.H., *An Essay in Aid of a Grammar of Assent*, 1870.
Newman, J.H., *Parochial and Plain Sermons*, 8 vols, 1898–1901, vol. VII, 1899, vol. VIII, 1899.
Newman, J.H., *Sermons Preached on Various Occasions*, 1900.
Newton, I., *Philosophiæ Naturalis Principia Mathematica*, 1687.
Newton, I., *Correspondence*, ed. H.W. Turnbull, J.F. Scott, A.R. Hall and L. Tilling, 7 vols, 1959–77, vol. II, 1960.
Newton, I., *Mathematical Papers*, ed. D.T. Whiteside, 8 vols, 1967–81, vol. III, 1969; vol. VII, 1976.
Nicole, P. [pseud. Sieur de Chanteresne], *De l'Éducation d'un Prince, divisée en trois parties, dont la dernière contient divers traités utiles à tout le monde*, 1670.
Nicole, P., *Essais de Morale contenus en divers traités sur plusieurs devoirs importants*, 13 vols, 1730–5, vol. VIII, 1733.
von Nietzsche, F.W., *Werke*, 8 vols, 1930–32, vol. III, 1930, *Daybreak* [1881]; vol. VI, 1930, *The Will to Power* [1887]; vol. VIII, 1932, correspondence.
Noël, É., *Le Plein du Vide*, 1648.
Noël, É., *Plenum Experimentis Novis Confirmatum*, 1648.
Norman, B., 'Thought and Language in Pascal', *Yale French Studies*, November 1973, pp. 110–19.
*Norman, B., *Portraits of Thought, Knowledge, Methods, and Styles in Pascal*, 1988.
Nouet, J., *Réponses aux* Lettres provinciales *publiées par le secrétaire de Port-Royal contre les PP. de la Compagnie de Jésus, sur le sujet de la morale des dits Pères*, 1656–7.
Novalis, *Werke, Briefe, Dokumente*, 4 vols, 1953–7, vol. II, 1957.
Ochino, B., *Sermones*, 1544.
Ojardias, A., 'Divers Portraits de Pascal et des siens', *Gazette des Beaux-Arts*, September 1910, pp. 195–207.
Oldenburg, H., *Correspondence*, 13 vols, 1965–86, vol. I, 1965 [1641–62], vol. X, 1975 [1673–4].
Onfray, R., *L'Abîme de Pascal*, 1949.
Orcibal, J., *Les Origines du jansénisme*, 5 vols, 1947–62.

Ore, O., 'Pascal and the Invention of Probability Theory', *American Mathematical Monthly*, May 1960, pp. 409–19.
Oughtred, W., *Arithmeticæ in Numeris et Speciebus Institutio . . . quasi Clavis Mathematicæ est*, 1631.
Pacioli, L., *Summa de Arithmetica, geometria, proportioni et proportionalita*, 1963 [1494].
Palacios, M., *Los Precedentes musulmanes del Pari de Pascal*, 1920.
Pappus, *Mathematicæ Collectiones*, 1588 [320 A.D.].
Pariente, J.-C., *L'Analyse du Langage à Port-Royal: Six Études logico-grammaticales*, 1985.
*Parish, R. J., *Pascal's* Lettres provinciales*: A Study in Polemic*, 1989.
Pater, W. H., 'Pascal', *Contemporary Review*, February 1895, pp. 168–81.
Patrick, D. G. M., *Pascal and Kierkegaard: A Study in the Strategy of Evangelism*, 2 vols, 1947.
Payen, J., 'Les Exemplaires conservés de la machine de Pascal', *Revue d'Histoire des sciences et de leurs applications*, April–June 1963, pp. 161–78.
Payen, J., 'La Machine arithmétique de Pascal: à Propos de l'exemplaire signé et daté de 1652', *Revue d'Histoire des sciences et de leurs applications*, October–December 1965, pp. 389–91.
Peletier, J., *Les Six premiers livres des Éléments géométriques d'Euclide, avec les démonstrations de Jacques Peletier du Mans*, 1611 [1557].
Périer, É., preface, *Pensées de M. Pascal sur la Religion et sur quelques autres sujets*, 1670.
Perrault, C., *Les Hommes illustres qui ont paru en France pendant ce siècle, avec leurs portraits au naturel*, 2 vols, 1697–1700.
Perrault, C., *Mémoires de ma Vie* [by Charles Perrault, 1755], *Voyage à Bordeaux (1669)* [by Claude Perrault], ed. P. Bonnefon, 1909.
Peverone, G. F., *Due Brevi e Facili Trattati, il primo d'Arithmetica; l'altro di Geometria*, 1558.
Philosophical Transactions [of the Royal Society], 1665–
*Picard, É., 'Pascal mathématicien et physicien', *Revue de France*, 15 July 1923, pp. 272–83.
Pintard, R., 'Pascal et les libertins', *Pascal présent, 1662–1962*, 1962, pp. 105–30.
Pinthereau, F., *Les Impostures et les Ignorances du libelle intitulé* la Théologie morale des jésuites, 1644.
Pirot, G., *Apologie pour les Casuistes contre les calomnies des jansénistes*, 1657.
Plato, *Portrait of Socrates*, ed. R.W. Livingstone, 1938, pp. 73–198, 'Phædo' [*circa* 380 B.C.].
Pommier, J.-M., *La Vie de Pascal, par Gilberte Périer. Étude critique et littéraire*, 1964.
Pomponazzi, P., *De Fato, Libero Arbitrio et Prædestinatione*, 1567 [but secretly written in 1520].
Poncelet, J.-V., *Traité des propriétés projectives des figures*, 1822.
Popper, K. R., *The Logic of Scientific Discovery*, 1959.
Port-Royal: *The Port-Royal Art of Thinking*, The First Class of the Course of Education Pursued at the Universities of Cambridge and Oxford, 1818.
Potts, D. C., 'Pascal's Contemporaries and *le Divertissement*', *Modern Language Review*, January 1962, pp. 31–40.

Prothero, R. E., *Private Letters of Edward Gibbon (1753–1794)*, ed. R. E. Prothero, 2 vols, 1896.
Pucelle, J., 'La Dialectique du Renversement du pour au contre et l'antithétique pascalienne', *Méthodes chez Pascal*, 1979, pp. 445–61.
*Pugh, A. R., *The Composition of Pascal's Apologia*, 1984.
Rabelais, F., *Œuvres. Gargantua et Pantagruel*, 5 vols, 1573.
Racine, J., *Œuvres complètes*, 2 vols, 1950–52.
Rauh, F., 'La Philosophie de Pascal', *Revue de Métaphysique et de morale*, April–June 1923, pp. 307–44.
Reginaldus [i.e., Regnault], V., *Praxis Fori Pænitentialis ad Directionem Confessarii*, 1620.
Réguron, P., *Les Origines du mouvement anti-janséniste et l'évolution de Pascal des* Provinciales *aux* Pensées, 1934.
Reisler, M., '*Persuasion through Antithesis*, An Analysis of the Dominant Rhetorical Structure of Pascal's *Lettres provinciales*', *Romanic Review*, May 1978, pp. 172–85.
Rémond de Montmort, P., *Essai d'Analyse sur les jeux de hasard*, 1708.
Renan, J.-E., *Correspondance, 1846–1871*, 1926.
Renan, J.-E., *Œuvres complètes*, 10 vols, 1947–61, vol. I, pp. 9–319, *Questions contemporaines* [1868], pp. 283–307, 'Du Libéralisme clérical' [written in 1848].
Renouvier, C.-B., *Philosophie analytique de l'histoire. Les Idées, les religions, les systèmes*, 4 vols, 1896–7, vol. IV, pp. 65–78, 'Les *Pensées* de Pascal au XIXe siècle'.
Renouvier, C.-B., *Histoire et solution des problèmes métaphysiques*, 1901.
Rescher, N., *Pascal's Wager: A Study of Practical Reasoning in Philosophical Theology*, 1985.
Rex, W., *Pascal's* Provincial Letters*: An Introduction*, 1977.
Roberts, D. E., *Existentialism and Religious Belief*, 1957.
de Robillard de Beaurepaire, C.-M., *Blaise Pascal et sa famille à Rouen de 1640 à 1647*, 1902.
Rochot, B., 'Comment Gassendi interprétait l'expérience du Puy de Dôme', *Revue d'Histoire des sciences et de leurs applications*, January-March 1963, pp. 53–76.
Rohault, J., *Traité de Physique*, 3rd edn, 2 vols, 1676.
Rouse Ball, W. W., *Mathematical Recreations and Essays*, 9th edn, 1920.
Russell, B. A. W., *A History of Western Philosophy*, 1945.
Russell, O. W., *Humor in Pascal*, 1977.
Russier, J., *La Foi selon Pascal*, 2 vols, 1949.
Russo, F., 'Pascal et l'analyse infinitésimale', *Revue d'Histoire des sciences et de leurs applications*, July–December 1962, pp. 303–20.
van Ruysbroeck, J., *Selected Works*, ed. E. Hello, 1912.
*Ryan, J.K., 'The Argument of the Wager in Pascal and Others', *The New Scholasticism*, July 1945, pp. 233–50.
de Sá, M., *Aphorismi Confessariorum ex Doctorum Sententiis collecti*, 1618 [1595].
*Sadoun-Goupil, M., 'L'Œuvre de Pascal et la physique moderne', *Revue d'Histoire des sciences et de leurs applications*, January–March 1963, pp. 23–52.
Saint-Ange, *see* Forton, J. (1637–45).

de Saint-Évremond, C.-M. de St-D., *Œuvres mêlées*, 1680, pp. 1–7, 'Pensées sur l'honnêteté', pp. 8–12, 'Description de l'honnête homme' [by D. Mitton].
Sainte-Beuve, C.-A., *Port-Royal*, 5 vols, 1840–61, vol. III, 1848, vol. IV, 1859.
Sainte-Beuve, C.-A., *Portraits contemporains*, rev. edn, 3 vols, 1855, vol. III, pp. 306–29, '*Pensées, Fragments et lettres* de Blaise Pascal'.
Sánchez, T., *Opus Morale in Præcepta Decalogi, sive Summa Casuum Conscientiæ*, 2 vols, 1615 [1613].
Schärer, K., 'Pascal und das Problem der Sprache', *Romanische Forschungen*, 1980, pp. 74–87.
von Schlegel, K. W. F., *Geschichte der alten und neuen Literatur*, 1815.
*Schmitz du Moulin, H., *Blaise Pascal: Une Biographie spirituelle*, 1982.
Scott, J. F., *The Mathematical Work of John Wallis, D.D., F.R.S. (1616–1703)*, 1938.
Scott, J. F., *The Scientific Work of René Descartes, 1596–1650*, 1952.
Sedgwick, A., *Jansenism in Seventeenth-Century France: Voices from the Wilderness*, 1977.
*Sellier, P., *Pascal et la liturgie*, 1966.
*Sellier, P., *Pascal et saint Augustin*, 1970.
de Sévigné, M., *Correspondance*, 3 vols, 1963–78.
Sheynin, O. B., 'Early History of the Theory of Probability', *Archive for History of Exact Sciences*, August 1977, pp. 201–59.
*Shiokawa, T., *Pascal et les Miracles*, 1977.
de Silhon, J., *De l'Immortalité de l'âme*, 1634.
Sirmond, A., *De Immortalitate Animæ, demonstratio physica et aristotelica, adversus Pomponatium et asseclas*, 1635.
Sirmond, A., *La Défense de la Vertu*, 1641.
Smart, R. N., *Historical Selections in the Philosophy of Religion*, ed. R.N. Smart, 1962.
Smart, R. N., *Philosophers and Religious Truth*, 1964.
Soria-Butron, J., *Epilogus Summarum, sive Amplissimum Compendium*, 1650. (No copy of a 1656 edition of this book is known to exist.)
Spink, J. S., *French Free-Thought from Gassendi to Voltaire*, 1960.
Spinoza, B., *Ethics*, Part V [1677].
Spinoza, B., *Correspondence*, ed. A. Wolf, 1928.
*Steinmann, J., *Pascal*, 2nd edn, 1962.
Stephen, L., *Studies of a Biographer*, 4 vols, 1898–1902, vol. II, 1898, pp. 241–84, 'Pascal'.
Stevin, S., *L'Arithmétique*, 1585.
Stevin, S., *De Beghinselen des Waterwichts*, 1586.
Stewart, H. F., *The Holiness of Pascal*, 1915.
Stewart, H.F., 'A Note on Pascal's *Pari*', *A Miscellany of Studies in Romance Languages & Literatures Presented to Leon E. Kastner*, 1932, pp. 481–8.
Stifel, M., *Arithmetica Integra*, 1544.
Strachey, G. L., *Landmarks in French Literature*, 1912.
Struik, D. J., *A Source Book in Mathematics, 1200–1800*, ed. D.J. Struik, 1969.
Suarès, A., *Puissances de Pascal*, 1923.
Suárez, F., *Tractatus de Legibus ac Deo Legislatore*, 1612.
Tacquet, A., *Cylindricorum et Annularium*, 1651.

Tallemant des Réaux, G., *Historiettes . . . , Mémoires pour servir à l'histoire du XVIIe siècle*, 2 vols, 1960–1 [1657–9].
Tanner, A., *Theologia Scolastica, Speculativa et Pratica*, 3 vols, 1621–7.
Tannery, P., *Mémoires scientifiques*, vol. VI, 1926, pp. 79–114, 269–79, 'Pascal et Lalouvère'.
Tartaglia, N. F., *General Trattato di Numeri et Misure*, 6 parts, 1556–60, vol. I, 1556.
Taton, R., *Le Calcul mécanique*, 1949.
Taton, R., *L'Œuvre mathématique de Desargues*, 1951.
*Taton, R., 'L'*Essay pour les Coniques* de Pascal', *Revue d'Histoire des sciences et de leurs applications*, January–March 1955, pp. 1–18.
*Taton, R., 'L'Œuvre de Pascal en géométrie projective', *Revue d'Histoire des sciences et de leurs applications*, July–December 1962, pp. 197–252 (1962a).
Taton, R., 'Pascal et la Machine arithmétique', *Pascal et Port-Royal*, 1962, pp. 67–72 (1962b).
Taton, R., 'Sur l'Invention de la machine arithmétique', *Revue d'Histoire des sciences et de leurs applications*, April–June 1963, pp. 139–60.
*Taton, R., 'Pascal, Blaise', *Dictionary of Scientific Biography*, vol. X, 1974, pp. 330–42.
Taveneaux, R., *La Vie quotidienne des jansénistes aux XVIIe et XVIIIe siècles*, 1973.
Tertullian, *De Carne Christi* [*circa* 206 A.D.].
*Thirouin, L., *Le Hasard et les Règles: le modèle du jeu dans la pensée de Pascal*, 1991.
St Thomas Aquinas, *Summa Contra Gentiles*, 4 vols, 1923–9 [1258–9].
Thurot, F.-C.-E., 'Recherches historiques sur le principe d'Archimède (*suite et fin*)', *Revue archéologique*, July 1869, pp. 14–23.
Todhunter, I., *A History of the Mathematical Theory of Probability from the Time of Pascal to that of Laplace*, 1865.
de Toledo [i.e., Toletus], F., *Summa Casuum Conscientiæ*, 1610.
*Topliss, P. M., *The Rhetoric of Pascal. A Study of his Art of Persuasion in the* Provinciales *and the* Pensées, 1966.
Torricelli, E., *Opera Geometrica*, 2 parts, 1644, part II, pp. 85–90, 'Appendix de Dimensione Cycloidis', part II, pp. 91–2, 'Scholium de Cycloidibus Aliarum Specierum'.
Turnbull, H. W., *The Great Mathematicians*, 1929.
Turner, A.J., *Early Scientific Instruments, 1400–1800*, 1987.
Turquet-Milnes, G. R., *From Pascal to Proust. Studies in the Genealogy of a Philosophy*, 1926, pp. 29–59, 'Bergson and Pascal'.
Turrianus [i.e., de Torres], L., *Selectæ Disputationes in Theologiam*, 1634.
Uhlír, A., 'Montaigne et Pascal', *Revue d'Histoire littéraire de la France*, July–September 1907, pp. 442–54.
de Valencia, G., *Commentariorum Theologicorum Libri Quatuor*, 5 vols, 1591.
Vacca, G. 'Sopra un probabile errore di Gabrio Piola', *Bollettino di Bibliografia e storia delle scienze matematiche*, January–March 1903, pp. 1–4.
Valensin, A., 'Note sur le Pari de Pascal', *Revue pratique d'apologétique*, 15 October 1919, pp. 65–74.
Valerio, L., *De Centro Gravitatis Solidorum Libri Tres*, 1604.
Valéry, P., 'Variation sur une *Pensée*', *Revue hebdomadaire*, 14 July 1923, pp. 161–70.

Valot, S., *Regardons vivre Blaise Pascal*, 1945.
Vamos, M., 'Pascal's *Pensées* and the Enlightenment: the Roots of a Misunderstanding', *Studies on Voltaire and the Eighteenth Century*, vol. XCVII, 1972, pp. 7–145.
Vasquez, G., *Opuscula Moralia*, 1618.
Vaughan, R. A., *Hours with the Mystics*, 2 vols, 1856.
de Vauvenargues, L. de C., *Introduction à la connaissance de l'esprit humain, suivie de réflexions et de maximes*, 1746.
Viète, F., *Apollonius Gallus, seu Exsuscitata Apollonii Pergæi* Περι επαθῶν *Geometria* [*The French Apollonius, or New Life Breathed into the Geometry of Apollonius of Perga* Concerning Contacts], 1600.
Vincent de Beauvais, *Bibliotheca Mundi, seu Speculum Naturale, Doctrinale, Morale, Historiale*, 4 vols, 1624 [*circa* 1260].
Voisin, J., 'Leibniz et le calcul intégral', *Revue des Questions scientifiques*, April 1967, pp. 233–46.
Voltaire, *Lettres Philosophiques*, 1734.
Voltaire, *Le Siècle de Louis XIV*, 2 vols, 1751.
Voltaire, *Œuvres complètes*, ed. A. de Condorcet, 70 vols, 1785–9, vol. LI, 1785.
Voltaire, *Œuvres complètes*, vol. CXVIII, 1974.
Wallis, J., *Grammatica Linguæ Anglicanæ*, 1653.
Wallis, J., *Arithmetica Infinitorum*, 1655.
Wallis, J., *Mathesis Universalis, in Operum Mathematicorum pars prima/altera*, 1657–6.
Wallis, J., *Tractatus duo, prior de Cycloide et Corporibus inde genitis, posterior epistolaris in qua agitur de Cissoide et Corporibus inde genitis*, 1659.
Ward, S., Ward, John [otherwise Seth], *The Lives of the Professors of Gresham College, London*, 1740.
Waterman, M., *Voltaire, Pascal and Human Destiny*, 1942.
Watson, J. D., *The Double Helix: A Personal Account of the Discovery of the Structure of DNA*, 1968.
Webb, C. C. J., *Pascal's Philosophy of Religion*, 1929.
Wells, A. N., *Pascal's Recovery of Man's Wholeness*, 1965.
Westfall, R. S., *The Life of Isaac Newton*, 1993.
Wetsel, D., L'Écriture et le reste: *The* Pensées *of Pascal in the Exegetical Tradition of Port-Royal*, 1981.
Whiteside, D. T., 'Wren the Mathematician', *The Royal Society, its Origins and Founders*, ed. H. B. Hartley, 1960, pp. 107–11.
*Whiteside, D. T., 'Patterns of Mathematical Thought in the Later Seventeenth Century', *Archive for History of Exact Sciences*, July 1961, pp. 179–388.
Wren, C., *Parentalia, or Memoirs of the Family of the Wrens*, ed. S. Wren, 1750.
Wright, C. J., *Miracle in History and in Modern Thought*, 1930.
Yhap, J., *The Rehabilitation of the Body as a Means of Knowing in Pascal's Philosophy of Experience*, 1991.

Louis Lafuma's edition of Pascal's *Œuvres complètes*, published in 1963, is a valuable adjunct to research.

A List of Pascal's Writings

Below is a list of those writings of Pascal – authorial, reported, summarized, contributed, envisaged or, just occasionally perhaps, supposititious – to which this book refers.

Titles of works projected by him but never written or else subsequently lost are indicated in roman type. Whether some of these works were ever written, in whole or in part, is a moot point.

Works by or attributed to Pascal, extant in whole or in substantial part, are indicated in *italics*. Works published under his own name during his lifetime are listed in bold italics.

Dates within brackets denote the known or conjectured year of writing.

The key words by which titles are referred to in the Index are designated in SMALL CAPITALS.

The definite and indefinite articles are omitted as the first words of English titles.

This list disregards most of the letters written by Pascal and certain ancillary writings.

ADDRESS TO THE Most Celebrated PARISIAN ACADEMIES of the Mathematical Sciences: Celeberrimæ Matheseos Academiæ Parisiensi (1654).[1]

On the ART OF PERSUASION: *De l'Art de persuader* (*circa* 1657).[2]

AT PORT-ROYAL: *A Port-Royal* (*A.P.R.*).[3]

COMPARISON BETWEEN CHRISTIANS *of Early Times and Those of Today*: *Comparaison des Chrétiens des premiers temps avec ceux d'aujourd'hui* (*circa* 1655).

COMPLETE WORK ON CONIC SECTIONS: Conicorum Opus Completum (1648–54).[4]

CONICAL CONTACTS: Tactiones Conicæ (1653).[5]

---

[1] The Address to the Most Celebrated Parisian Academies of the Mathematical Sciences, lost since 1697 (**II 220**), was lent to Leibniz early in June 1675; he copied it out in full (Hofmann, J. E. (1974), 179). The first *Recueil Guerrier* (xxv) also records this 'Address'.

[2] *On the Art of Persuasion* and *On the Geometrical Mind* are essentially one text. As such, they are referred to in Arnauld, A. and Nicole, P. (1664), 15 as 'a short unpublished essay, written by the late M. Pascal and entitled by him *On the Geometrical Mind*'. Pascal is not mentioned by name in the 1st edn (Arnauld, A. and Nicole, P. (1662), 18).

[3] *Thought* 149*. This is the outline of a talk about the *Thoughts* given by Pascal at Port-Royal des Champs, probably in May 1658.

[4] Contents summarized by Tschirnhaus and Leibniz. Includes The Generation of Conic Sections, completed in March 1648 and also known as the Treatise on Conic Sections.

[5] See Chapter II n. 31. Leibniz has left some clue as to the content of this only partially completed work (64). See also **II 225** and **III 301–2**.

CONVERSATION OF M. PASCAL AND M. DE SACI *Concerning Epictetus and Montaigne: Entretien de M. Pascal et de M. de Saci sur Épictète et Montaigne* (1655).[6]

On the CONVERSION OF THE SINNER: *Sur la Conversion du pécheur* (1655).

DEPOSITION *on Behalf of the Parish Priests of Paris Against the Book entitled* Defence of the Casuists: *Factum pour les Curés de Paris contre le livre intitulé* Apologie pour les casuistes (1658).

DISCOURSE ON THE PASSIONS OF LOVE: *Discours sur les Passions de l'amour* (*circa* 1653–4).[7]

DISCOURSES ON THE WORLDLY CONDITION OF THE GREAT: *Discours sur la Condition des grands* (1660).[8]

DRAFT MANDAMUS *Against the* Defence of the Casuists: *Projet de Mandement contre l'*Apologie des casuistes (1658).

ELEMENTARY GEOMETRY: Éléments de géométrie (*circa* 1657).[9]

ESSAY ON CONIC SECTIONS: ***Essai pour les Coniques*** (1640).[10]

FIFTH STATEMENT *by the Parish Priests of Paris Against the Book entitled* Defence of the Casuists: *Cinquième Écrit des Curés de Paris contre le livre intitulé* Apologie pour les casuistes (1658).

FIRST CIRCULAR LETTER *Concerning the Cycloid (Problems Concerning the Cycloid, Set in June 1658: De Cycloïde. Problemata de Cycloïde, Proposita Mense Junii 1658)* (1658).

FIRST MANDAMUS *of the Vicars General Concerning the Signing of the Formulary Condemning* De Jure *Jansen's Five Propositions: Premier Mandement des vicaires généraux sur la signature du Formulaire, portant condamnation de droit des Cinq Propositions de Jansénius* (1661).[11]

FOOTNOTE TO THE SEQUEL TO THE HISTORY OF THE CYCLOID: *Addition à la Suite de l'Histoire de la roulette* (1659).[12]

FRENCH APOLLONIUS[13] Improved Upon: Promotus Apollonius Gallus (1654).[14]

GENERAL TREATISE ON THE CYCLOID: *Traité Général de la roulette* (1658–9).

---

[6] Desmolets, P.-N. (1728), V(2), 237–70. The conversation was recorded by Saci's secretary Nicolas Fontaine.

[7] This is most probably a cento embodying some utterances of Pascal.

[8] First published by Nicole, P. (1670), 272–85.

[9] This inspired Arnauld, A. (1667).

[10] Leibniz and his friend Tschirnhaus were lent this printed scientific paper in Paris in January 1676 (Hofmann, J. E. (1974), 179–80; Aiton, E. J. (1985), 56). Together with all Pascal's other mathematical manuscripts borrowed by Leibniz from étienne, Louis and Blaise Périer (and returned to Pascal's nephews on 30 August 1676), it had vanished by the time of the Abbé Gilles Filleau des Billettes's letter to Leibniz dated 28 May 1697 (II 220). See also Chapter II n. 1.

[11] This was published on 8 June 1661.

[12] The *Footnote to the Sequel to the History of the Cycloid* was published on 20 January 1659.

[13] François Viète (1540–1603).

[14] See Chapter II n. 34. This tract would have been a refinement of the work of François Viète, who succeeded in reconstituting Apollonius's work on contacts. In the event the application of Viète's findings to the problem of spherical contacts was carried out by Fermat (de Fermat, P. (1891–1922), I 52–69; French translation: III 49–66).

GENERATION OF CONIC SECTIONS: Generatio Conisectionum (1648[–54]).[15]
On the GEOMETRICAL MIND: *De l'Esprit géométrique* (*circa* 1657).[2]
GEOMETRY OF CHANCE: De Aleæ Geometria (1654).[17]
GEOMETRY OF CHANCE: La Règle des Partis: De Aleæ Geometria (1654).[17]
HISTORY OF THE Roulette, *otherwise known as the Trochoid, or* CYCLOID, *in which it is told by What Steps people have arrived at a Knowledge of the Nature of that Line: Histoire de la roulette, appelée autrement la trochoïde, ou la cycloïde, où l'on rapporte par quels degrés on est arrivé à la connaissance de la nature de cette ligne* (1658).[18]
HISTORY OF THE TROCHOID, or Cycloid, in French: la roulette, *in which it is told by What Steps people have arrived at an intimate Knowledge of the Nature of that Line: Historia Trochoïdis, sive Cycloïdis, gallice:* la Roulette, *in qua narratur Quibus Gradibus ad Intimam illius Lineæ Naturam Cognoscendam Perventum sit* (1658).[18]
INTRODUCTION TO GEOMETRY: Introduction à la géométrie (*circa* 1657).[19]
LETTER to M. and Mme Périer . . . ON the Occasion of THE DEATH OF M. PASCAL senior: Lettre à M. et Mme Perier . . . à l'occasion de la mort de M. Pascal le père (1651).
LETTER from A. Dettonville TO MONSIEUR A.D.D.S.:[20] *Lettre de A. Dettonville à Monsieur A.D.D.S.* (1658)
LETTER from Pascal to CARCAVI: Lettre de Pascal à Carcavi (1659).
LETTER from M. Dettonville to M. HUYGENS concerning the Dimensions of the Lines of All Cycloids: Lettre de M. Dettonville à M. Huyghens sur la dimension des lignes de toutes les roulettes (1659).
Letter from M. Dettonville TO M. de SLUZE, concerning the Staircase, Cylindrical Triangles, and the Solid of Revolution Generated by a Spiral Rotating Around a Cone: Lettre de M. Dettonville à M. de Sluze, de l'Escalier, des triangles cylindriques, et de la spirale autour d'un cône (1658).
LETTERS TO MLLE DE ROANNEZ: Lettres à Mlle de Roannez (1656).

---

15 This propounds the theorem of the Mystic Hexagon (i.e. Pascal's Theorem), discovered in 1639 but merely heralded rather than formally published during Pascal's lifetime. Its contents have been summarized by Tschirnhaus and Leibniz, who were shown it in 1676 (**II 220–4**). By 28 May 1697 The Generation of Conic Sections, otherwise known as the Treatise on Conic Sections, had disappeared (**II 220**). It was no longer in Louis Périer's possession at his death in 1713. Leibniz's notes on Pascal's tract on conic sections, together with Tschirnhaus's copy of the figure of the Mystic Hexagon and The Generation of Conic Sections, have been published in **II 217–43**. See also 66–70, 1382–7.
16 This would perhaps have become the overall title of a general work on the probability calculus which Pascal never managed to put together.
17 This consists partly of *The Probability Calculus* and partly of the *Treatise on the Arithmetical Triangle*, with its two annexes *Multiple Numbers* and *The Summing of the Powers of Integers*. The *Treatise on the Arithmetical Triangle* was probably written in August 1654.
18 Editions of the *History of the Cycloid* were published simultaneously in Latin and in French on 10 October 1658.
19 Contents summarized by Leibniz. In June 1675 Leibniz was shown this document by the Abbé Gilles Filleau des Billettes, who allowed him to make excerpts from it (Hofmann, J. E. (1974), 179).
20 'Monsieur A.D.D.S.' very probably denotes Antoine Arnauld: 'Monsieur Arnauld, Docteur de Sorbonne'

LETTERS from A. Dettonville TO M. de CARCAVI: Lettres de A. Dettonville à M. de Carcavi (1658–9).[21]
On MAGIC NUMBERS: De Numeris Magico Magicis (1654).[22]
MEMORIAL (913*): *Mémorial* (1654).[23]
METHOD OF PERSPECTIVES: Perspectivæ Methodus (*circa* 1654).[17]
Concerning the Recognition of MULTIPLE NUMBERS by [Means of the Divisibility of] *the Mere Sum of their Digits: De Numeris Multiplicibus ex Sola Characterum Numericorum Additione Agnoscendis* (1654).[17]
MYSTERY OF JESUS (919*): *Le Mystère de Jésus* (1655).
NARRATIVE ACCOUNT OF THE GREAT EXPERIMENT on the Equilibrium of Liquids: Récit de la Grande expérience de l'équilibre des liqueurs (1648).[25]
NEW EXPERIMENTS Concerning Vacuums: Expériences nouvelles touchant le vide (1647).[26]
Fragment of a NINETEENTH PROVINCIAL LETTER: Fragment d'une Dix-neuvième Lettre provinciale (1657).
PLANE LOCI: Loci Plani (1653–4).[27]
PORT-ROYAL GRAMMAR: Grammaire de Port-Royal (1655–7).[28]
PORT-ROYAL LOGIC: Logique de Port-Royal (*circa* 1657).[28]
PRAYER to God CONCERNING THE PROPER USE OF ILLNESSES: Prière pour demander à Dieu le bon usage des maladies (1659).[30]
PROBABILITY CALCULUS: Règle des Partis (1654[31]).[17]

---

[21] Pascal's all-important work on the cycloid was pseudonymously published in the form of four letters (10 December 1658–January 1659) purporting to have been written by Amos Dettonville. The prize questions on the cycloid (the three *Circular Letters*) had also been issued in Dettonville's name. The authorship of the *Letters to Carcavi* was clear, however, to Ismaël Boulliau and to most of Pascal's intimates.
[22] Antoine Arnauld has left clear indications of the content of this work, a large portion of which seems to have been completed: Arnauld, A. (1667), 325–45.
[23] The earliest components of the *Thoughts* consist essentially of the *Memorial* (23–4 November 1654), the *Mystery of Jesus* (*circa* January 1655) and the section on miracles (830*-912*: some from about September 1656, but chiefly from January-February 1657). Many of the *Thoughts* date from the spring of 1657. Some (e.g., 920*, 952*, 956*) plainly date from the time of the preliminary drafting of the Nineteenth Provincial Letter in that year. The main effort was almost certainly over by 1658.
[24] This would have carried a stage further the work on the science of perspective done by Gérard Desargues.
[25] This narrates the experiment conducted on the Puy de Dôme on 19 September 1648.
[26] *New Experiments Concerning Vacuums* was published on 8 October 1647.
[27] This would have been a reconstitution of Apollonius of Perga's lost treatise on loci that are either circles or straight lines. Apollonius's *Plane Loci* was, in fact, reconstituted by Fermat (de Fermat, P. (1891–1922), I 3–51; French translation: III 3–48).
[28] Pascal contributed (234 n. 32, 267) to Lancelot, C. and Arnauld, A. (1660).
[29] Pascal contributed (234 n. 35, 258) to Arnauld, A. and Nicole, P. (1662)/Arnauld, A. and Nicole, P. (1664).
[30] Perhaps composed in June 1659.
[31] Exchange of letters between Pascal and Fermat, June-July (**i**), 29 July (**ii**), 24 (**iii**) and 29 August (**iv**) and 25 September 1654 (**v**). Pascal was the author of letters **ii** and **iii**. The correspondence between the two men was not published until 1679 (**ii**, **iii**) and 1779 (**i**, **iv**, **v**): de Fermat, P. (1891–1922), II 288–314. Fermat's **iv** was despatched before he received **iii**, to which he replied in **v**.

PROVINCIAL LETTERS: *Lettres provinciales* (1656–7).[32]
SECOND CIRCULAR LETTER *Concerning the Cycloid (Concerning the Cycloid. An Addition to That Argument: De Cycloïde. De eodem Argumento Additamentum)* (1658).
SECOND STATEMENT *by the Parish Priests of Paris Against the Book entitled* Defence of the Casuists: *Second Écrit des Curés de Paris contre le livre intitulé* Apologie pour les casuistes (1658).[33]
SEQUEL TO THE HISTORY OF THE CYCLOID, *in which are seen the Procedures of a Person who had tried to claim Credit for the Solution of the Problems Set upon this Subject: Suite de l'Histoire de la roulette, où l'on voit le procédé d'une personne qui s'était voulu attribuer l'invention des problèmes proposés sur ce sujet* (1658).[34]
SIXTH STATEMENT *by the Parish Priests of Paris Against the Book entitled* Defence of the Casuists: *Sixième Écrit des Curés de Paris contre le livre intitulé* Apologie pour les casuistes (1658).
SOLID LOCI: Loci Solidi (1653).[35]
SPHERICAL CONTACTS: Tactiones Sphæricæ (1654).[14]
STATEMENT ON THE SIGNING OF THE FORMULARY: *Écrit sur la signature du Formulaire* (1661).[36]
SUMMARY OF THE LIFE OF JESUS CHRIST: *Abrégé de la Vie de Jésus-Christ* (1655).
SUMMING OF THE POWERS OF INTEGERS: *Potestatum Numericarum Summa* (1654).[17]
THIRD CIRCULAR LETTER *Concerning the Cycloid (Concerning the Cycloid. Notes on Certain Solutions to the Cycloid Problem: De Cycloïde. Annotata in quasdam Solutiones Problematum de Cycloïde)* (1658).[37]
THOUGHTS: *Pensées* (1654[23]–61).[38]
TREATISE ON CONIC SECTIONS: Tractatus Conisectionum (1648[-54]).[15]
TREATISE ON MECHANICS: Traité de Mécanique (1651).
TREATISE ON SOUNDS: Traité sur les Sons (1634).
TREATISE ON THE ARITHMETICAL TRIANGLE: *Traité du Triangle arithmétique* (1654).[17]
TREATISE ON THE EQUILIBRIUM OF LIQUIDS: *Traité de l'Équilibre des liqueurs* (*circa* 1653).

---

[32] I (23.1.1656), II (29.1.1656), III (9.2.1656), IV (25.2.1656), V (20.3.1656), VI (10.4.1656), VII (25.4.1656), VIII (28.5.1656), IX (3.7.1656), X (2.8.1656), XI (18.8.1656), XII (9.9.1656), XIII (30.9.1656), XIV (23.10.1656), XV (25.11.1656), XVI (4.12.1656), XVII (23.1.1657), XVIII (24.3.1657).
[33] Of the six *Statements by the Parish Priests of Paris* it would seem that three (II, V, VI) were by Pascal, one (III) by Arnauld and one (IV) by Nicole. See **VII 353**.
[34] The *Sequel to the History of the Cycloid* was published on 12 December 1658.
[35] See Chapter II n. 31. Leibniz has left some clue as to the content of this work (64–5), a work substantially completed but now lost.
[36] This is known only from a copy by Pierre Nicole (1075). It dates from November 1661.
[37] The *Third Circular Letter* was published on 9 October 1658.
[38] For fuller details concerning the protracted publication of the *Thoughts* (the Recueil d'Utrecht, the Recueils Guerrier etc) see Pascal: *Œuvres complètes*, Pléiade edn, 1954, xxiv–xxv.

TREATISE ON THE WEIGHT OF THE AIR MASS: Traité de la Pesanteur de la masse de l'air (*circa* 1654).

TREATISE ON VACUUMS: Traité du Vide (1651–4).[39]

TREATISES ON GEOMETRY: Traités de géométrie (1658–9).[40]

WRITINGS ON GRACE: Écrits sur la Grâce (1657–8).

[39] Much of this projected work remained unwritten.

[40] This is the collective name for four treatises on projective geometry.

Index